Food Science Text Seri

MW00610303

The Food Science Text Series provides faculty with the leading teaching tools. The Editorial Board has outlined the most appropriate and complete content for each food science course in a typical food science program and has identified textbooks of the highest quality, written by the leading food science educators.

For further volumes:
http://www.springer.com/series/5999

Harry T. Lawless

Laboratory Exercises for Sensory Evaluation

 Springer

Harry T. Lawless
Department of Food Science
Cornell University
Ithaca, NY, USA

ISSN 1572-0330
ISBN 978-1-4614-5682-7   ISBN 978-1-4614-5713-8 (eBook)
DOI 10.1007/978-1-4614-5713-8
Springer New York Heidelberg Dordrecht London

Library of Congress Control Number: 2012948339

Printed on acid-free paper

Springer is part of Springer Science+Business Media (www.springer.com)

Preface

Laboratory exercises are a necessary part of scientific education. They teach students how to perform various tests and procedures through learning by doing. It is of course possible to teach sensory evaluation only from lectures and book learning, but our philosophy is that a great understanding is achieved by hands-on experience. Laboratory exercises provide important opportunities to reinforce concepts, principles, and practices discussed in lectures. Furthermore, sensory evaluation is a science of test procedures, statistical analyses, and proper interpretation of results. The procedures, analysis, and conclusions are best learned by actually conducting the tests. Many of the critical details in methods and good practices are hard to absorb and remember without having been in the test situation itself.

The objectives of a laboratory course in sensory evaluation are:

1. To teach students how to set up and conduct sensory tests.
2. To give students experience in analyzing data from sensory tests.
3. To develop an understanding of the meaningful, valid interpretation of results.
4. To develop skills in writing reports and industrial memos.
5. To give students experience in participating in sensory tests.
6. To reinforce information taught in the lecture section of the course.
7. To provide experience in evaluating a variety of products using the senses.

Sensory evaluation is unlike chemistry and biology. Analytical procedures in those disciplines can be spelled out in exquisite detail and followed in cookbook fashion. However, every food product is different, and the real challenge is in devising a sensory test method that fits the product and is both sensitive and unbiased. This need for flexibility can be quite challenging and unsettling to people who are used to rote procedures. This was written to be procedurally specific and "cookbookish" as is reasonable. I have noted where procedures may be modified to suit the situation of individual schools and professors. I hope that each professor may use these exercises as a starting point and allow his or her own expertise to contribute modifications and discussion points as they see fit.

The structure of this manual includes eleven longer laboratory exercises, in Chaps. 3 through 13. They are best suited for a 3-h laboratory period but require less time if more preparation is done ahead of time. The first lab concerns panelist selection and testing the acuity of a person's senses of taste and smell. Three exercises are related to discrimination, thresholds, and signal detection theory. Two scaling exercises involve a comparison of methods and time-intensity scaling. Three exercises are related to descriptive analysis, including flavor profile, ballot and term generation, and use of reference standards. Finally, two labs are concerned with acceptance and preference testing and product optimization. There is one group project in Chap. 14 on descriptive analysis, a topic that lends itself to a group effort. There are four briefer exercises in Chap. 15, on probability, shelf life, consumer questionnaires, and development of a procedure for a difficult-to-handle meat product. The final chapter gives some examples of statistical problem sets that could be used as part of the laboratory course or main sensory course.

All of the laboratory exercises in the current work were conducted in one form or another for many years at the Cornell University Food Science Department course in sensory evaluation, with the exception of Chap. 11, use of reference standards. That exercise is based on the actual experience of the author in training descriptive panels or being a panelist myself, and so I feel confident that Chap. 11 can be executed as written or with only minor modifications. At Cornell, the class consisted mostly of senior (fourth year) undergraduates and some graduate students. If your course is given earlier in terms of the students' experience or year, you may wish to consider simplifying some of the activities.

This manual owes a double debt of gratitude to Professor Hildegarde Heymann of the University of California at Davis. Professor Heymann is familiar to many of her colleagues and students due to her tenure at the University of Missouri Columbia, where we first began the construction of a comprehensive sensory evaluation textbook, and then for her return to the Davis campus to assume a sensory leadership role in the program in viticulture and enology. Almost two decades before, Professor Heymann was instrumental in the compilation of a set of laboratory exercises in sensory evaluation, under the auspices of the Sensory Evaluation Division of the Institute of Food Technologists. Many of us contributed to that manual, and parallels can be seen in the lab exercises on scaling and acceptance/preference testing. It was intended as a guide to new instructors in sensory evaluation, as there was a dearth of professors in this field at the time (some would say there still is) and those of us "old hands" thought it would be great idea to share some of our labs that seemed to work well. That document was influential in the construction of this work. An even older document from Professor Pangborn's labs at Davis somehow found its way into my hands, I believe from the archives of Suzanne Pecore. Its level of detail was an eye-opener. Another important resource from my own experience was my undergraduate sensory psychology course, with labs from the scientists at the John B. Pierce Foundation, notably (in alphabetical order) Ellie Adair, Linda Bartoshuk, Bill Cain, Larry Marks, and Joe Stevens. I had many courses and labs in chemistry, but psychology labs with human data were very different.

Professor Heymann chose not to coauthor this current work, due to other commitments, but it was our intention early on to have a companion piece to Sensory Evaluation of Foods, Principles and Practices, and that it would be a kind of Lawless and Heymann, *part deux*. During various phases of this project, she offered invaluable guidance and commentary, as well as providing the author copies of her laboratory exercises for the sensory evaluation courses at Davis in viticulture and enology. In many ways, she should rightfully be a junior author.

Susan Safren and colleagues at Springer Publishing were patient editorial guides to the project. I appreciate her forbearance during a series of strategic changes during the construction and organization of this work. In particular, we suffered over whether there should be separate instructor and student sections for the labs, in totally separate parts of the book, or even whether there should be separate manuals. In the end, the current author decided (unilaterally) to merge the instructor and student sections for each lab and present each exercise as a kind of book chapter that could be purchased or downloaded separately. The negative aspect of that merger is that students may not be totally blinded to the nature of the samples, if they simply read ahead.

A debt of gratitude is due to my students and teaching assistants. Many of the TAs asked critical questions over the years that fettered out a lot of potential problems with the procedures. Others, thankfully few, manage to make serious but innocent mistakes. These two were informative. In order to avoid those potential mistakes, there are sections noted as "Notes and keys to successful execution" in each instructor section. One the whole, my TAs at Cornell were excellent in their effort, hard work, and attention to detail and taught me a lot. Pedagogy is a two-way street.

Professor Edgar Chambers IV was the source of the laboratory on the Flavor Profile Method, and I first encountered it during a workshop we gave at Kansas State University in the mid-1990s for new instructors in sensory evaluation. It was so much fun and so colorful that I incorporated it in a slightly modified form into my own course at Cornell the very next semester. Many thanks, Edgar. John Horne was my lab manager when most of these exercises were taking shape, and he applied his editorial hand to early versions. Kathy Chapman also served for many years in that support role.

Comments and feedback may be directed to the author at htl1@cornell.edu.

Ithaca, NY, USA   Harry T. Lawless

Contents

Part I

Introductory Material
and General Instructions

Introduction and General Instructions for Students

1.1 General Instructions for Students

Laboratory demonstrations and exercises will introduce and illustrate the principles and methods used to measure sensory attributes of foods and beverages. Attention must be given to the careful collection of data, statistical analysis, and interpretation of findings. After most of these laboratory exercises, students will decode and enter their own responses on master sheets. Lab reports may require different formats. Please check with your instructor or course website as to the format required for each report or if there is a default format.

Your enthusiastic participation will add to your enjoyment of the course as well as improving your sensory awareness. Make an effort to develop the following skills: careful attention to experimental procedures, a critical approach toward the reliability and validity of the results obtained, and a feel for the importance of professionalism in the reports, written and oral.

1.1.1 Guidelines for Participation

These laboratory tests are teaching demonstrations, not research experiments. You will often participate as both experimenter and judge, and therefore, you will frequently know the nature of the experiments in advance. You may have a tendency to try to "outguess" the tests. Please try to refrain from doing this, and remember, the important objective is often not to determine how good a "taster" you are but rather to teach you the appropriate procedures.

Please do not wear perfume, aftershave, or fragranced hand creams on laboratory days. If someone else can smell you, you may be denied access to the sensory laboratory. Do not bring food or beverages into the laboratory other than water. Please observe sanitary practices of hand washing before handling sample containers, etc.. Please try not to touch the drinking edge of glasses. If samples are given out for distribution "family style," there will be a serving spoon provided in addition to your individual tasting spoon. Do not confuse the two and do not use your individual tasting spoon for taking your sample from the communal vessel. At the conclusion of the class, please put garbage in the provided receptacles. Rinse dishes and spit cups or dispose of as indicated.

1.2 Options for Individual or Partner Work

1.2.1 General Option 1: Individual Work

1. Students must work individually.
2. Use one of the lab report formats indicated in the lab manual (or class website) for each lab exercise. If not specified, the standard report format will be used.

H.T. Lawless, *Laboratory Exercises for Sensory Evaluation*, Food Science Text Series 2,
DOI 10.1007/978-1-4614-5713-8_1, © Springer Science+Business Media New York 2013

3. Answer discussion questions briefly.
4. Cite references you used, including the specific section of your textbook if needed.
5. Reports must be typed.
6. Reports will be graded on content, organization, and demonstration of comprehension of material, language usage, and neatness. Check the course requirements and/or website before you start the reports. Lab reports may be graded on a five- or ten-point basis, depending upon the difficulty level of the exercise and the amount of work required.

1.2.2   General Option 2: Partner Work

1. Groups of two students may work on the data entry, statistical analyses, and graphics for the reports. However, each student must submit a separate report. Partners may submit copies of each other's graphs, but each student must write his/her own interpretation of data and conclusions. Individuals are responsible for accuracy and neatness of all material. (If your partner does a good job, you get the credit too; if your partner is inaccurate or sloppy, you lose points too.) Always indicate your partner's name on the cover sheet of your reports.
2. If the results are consistent with the literature, please indicate this. If the results are inconsistent with your expectations or the literature, suggest reasons for the possibly spurious results or ways in which the experiment should be redone.
3. Answer discussion questions briefly.
4. Always cite the references you used. If you read a paper do not use information from it in writing the report, do not include it. Only cite references you actually read!
5. Reports must be typed.
6. The reports should be brief and concise but complete. Each report must follow the format given for that specific report. Reports may require different formats.
7. Reports will be graded on content, organization, and demonstration of comprehension of material, language usage, and neatness. Check the course requirements and/or website before

you start the reports. Lab reports may be graded on a five- or ten-point basis, depending upon the difficulty level of the exercise and the amount of work required.

1.3   Suggested Lab Report Formats

1.3.1   Specific Formats

The examples shown here are generic formats that should be tailored to the specific situation. Some of the information discussed above may not be available or not needed. In industry, upper level managers will rarely read more than one or two pages. Thus, with the memo format, you must ensure that you get the important information on the first page. Any discussion or comparisons to previous findings or existing literature, if needed, should be made within the body of the memo, usually in the conclusions or in the results, if appropriate. Make sure that you are following the correct format for the specific laboratory, as stated in your syllabus, schedule, or course website.

1.3.1.1 Option 1: Concise Lab Report Format

Reports must contain the following information:
1. Title and number of the lab report, your name and student ID number, and date submitted
2. Objectives of the exercise
3. Methods, concisely stated
4. Results, as required by the individual exercise:
   (a) Describe what happened in words.
   (b) Use any statistical tests to back up what you just described.
   (c) Graphs and tables, if needed or required. See guidelines for graphing.
5. Answers to discussion questions
6. Appendix: Calculations used in any statistical analyses done "by hand," i.e., without a statistical program
7. References

Note: It is not necessary to append tables of raw data.

1.3.1.2 Option 2: Long Report Format

This is written similar to a scientific journal article. These formal laboratory reports must contain, in the following order:

1. Laboratory number and title.
2. Name and date (include name of partner if working in pairs).
3. Summary: This section will have a maximum of 300 words and summarizes the objective(s), the results, and the conclusions. If you were writing this report in industry, this may be all that your manager will read; thus, this section should be clear and to the point.
4. A brief review of the literature using any assigned papers for the lab, other reading assignments, and the textbook, if required for the specific report. Limit to one paragraph.
5. A brief summary of the materials and methods used to prepare the samples and to do the test and the data analyses.
6. Results: This section contains figures, tables, and statistical calculations.
   (a) Figures—See also graphing suggestions and guidelines below.
      (i) Figures should have a number and a complete title or caption.
      (ii) Figures should be completely labeled (both axes identified and the number of judges or judgments specified).
      (iii) For each data point (if possible), include a measure of dispersion around the means such as standard deviation, standard error.
   (b) Tables (see guidelines).
   (c) Statistical analyses (see guidelines).
7. Discussion and interpretation of the results, with comparisons to the literature as discussed in 4. Begin by stating whether the results support or fail to support any original assumption(s). The interpretation should be written briefly and clearly so as to justify conclusions. Interpretation of data should be logical and organized.
8. Concise answers to specific questions (if any) posed in manual.
9. Write a brief statement of your personal conclusions, based on the experimental results.

Indicate any suggestions for improvement and any surprises…

10. References cited, in alphabetical order. Give complete citation; use the form used by the Journal of Food Science or other journal as indicated by your instructor.
11. Number the pages. Staple them together.
12. Proofread your report. Do not simply rely on your spell-checker.

1.3.2 Other Formats Used in Industry and Product Research and Development

Technical, industry memos should use one of the following formats. A sample report is also shown in the appendix of this section with a simple two-page format. All memos should be clear, concise, and accurate. In the subject heading, you define the subject of the memo. In the body of the memo, the headings help the reader understand the message conveyed. Memos are usually structured in two parts: the identifying information and the body.

1.3.2.1 Complete Memo Format

A. Identifying Information (Sample)

MEMO

TO: I. Newton, VP for Research and Development
FROM: A. Einstein, Sensory Division Manager
SUBJECT: Results of sensory tests on Chianti.
DATE: June 11, 2009

The second and any subsequent pages of the memo often have the following header information in the upper left-hand corner of each page: name of recipient, date of memo, and page number

B. Body of the Memo

The body of a memo is always single spaced and should follow the basic structure of the given format. The format gives the reader the same sense of direction that a full-scale outline does in a formal report. Use headings and lists to clarify the structure.

Objective: The objective should answer the question "Why is he/she telling me this?" The best objective statements are direct and concise.

Product description: Be concise and complete.

Conclusions and recommendations: This is the most important part of the memo to the reader. Conclusions are not the statistical results; they are the reasoned facts that you draw out of the results. Recommendations are courses of action, reasoned from the conclusions. In some companies, this is the top-line information, following the heading.

Results are usually summarized on the second page and form the core of the memo. Results help to clarify the basis for the conclusions. Depending on the situation, the summary may be one sentence or may be a series of longer technical paragraphs.

Most executives will only read the first page (and if you are lucky the second) of the memo; however, supporting information such as graphs can be attached or appended if they communicate an important message or illustrate the results and help justify the conclusions.

1.3.2.2 Executive Summary Format

An executive summary consists of the following information and it must fit on a single page:
1. Header information—To: From: Date: Reference (topic):
2. Objectives of the study/test/report.
3. Conclusions. Note that these are not statistical results but are reasoned judgments based on the results.
4. Recommendations. These are courses of action, not conclusions or results.

Additional materials may be appended as suggested by your instructor, if needed.

1.4   Guidelines for Reporting Statistical Results: Use of Statistical Tables

1. A maximum of three figures beyond the decimal is sufficient for obtained statistics such as F-values, t-values, and correlation coefficients,

e.g., $F = 7.962$, $t = 2.356$, and $r = 0.581$. Your statistical program, if used, may give you more, but they should be rounded.

For sensory data, usually, only one figure beyond the decimal is generally reported, e.g., 6.7, 9.1, and 42.6 % (please use an appropriate way to round the data). In general, the number of figures used should reflect the standard error values. For example, if the standard error of the mean is 0.6, the means should not be reported to more than one decimal place. If the standard error were 0.006, that would suggest three points beyond the decimal was appropriate (this level of precision would be rare).

2. Probability: You must use the significance level that was predetermined. In most courses that is 5 % unless stated otherwise. If you have an exact probability value, you may give it.

As an aside: If a value is significant at $P < 0.001$, it is obviously also significant at $P < 0.01$ and $P < 0.05$. If using some indicator like an asterisk, you should only indicate one asterisk since we predetermined the alpha value at 5 %. The lower probability levels only indicate the likelihood of the result under a true null, which you are rejecting anyway. They do not indicate the strength of the affect, the likelihood of the difference (1 minus p), or anything else. Do not get in the bad habit of using multiple asterisks as permitted (unfortunately and erroneously) in some journals.

3. Be concise, specific, and clear when discussing statistical results in text. State what direction the difference or result was in, then use the statistics to back up your statement. Here are some examples:

The samples with increasing levels of sucrose were increasingly sweeter. The samples differed significantly ($F = 21.57$, 3/42 df, $p < 0.05$) and therefore…

The ascending series gave a higher value than the descending series ($t = 9.035$, 14 df, $p < 0.05$).

There was a positive correlation ($r = +0.972$, 32 df, $p < 0.05$) between X and Y. As X increased, Y also increased.

Table 1.1 Analysis of variance for perceived astringency of Cabernet Sauvignon wines

Source	Degrees of freedom	Sums of squares	Mean squares	F-value and level of significance
Judges	4	12.311	3.078	6.6*
Wines	2	90.844	45.422	96.8*
Subsamples	2	0.311	0.156	0.3
Judge* subsamples	8	4.356	0.544	1.2
Judge* products	8	21.822	2.728	5.8*
Subsamples* products	4	2.489	0.622	1.3
Error	16	7.511	0.469	

*is significant at $P < 0.05$

4. Do not use computer printouts "as is" for your tables. Do not simply cut and paste, but make a table in a format that is in a publishable condition. See guidelines for tables.
5. Reporting analysis of variance (ANOVA) results:
   (a) Create an ANOVA table.
   (b) Give it a table number and title. It should be possible to read only the title, the body of the table, and any footnotes to completely understand the table.
   (c) Identify the sources of variation (main effects: judges, reps, interaction effects, etc.).
   (d) Indicate significance with an asterisk (*) and identify the level (e.g., $p < .05$) in a footnote. If available, use the exact p-values.
   (e) Calculate Fisher's LSD (or Tukey HSD, Duncan tests, or similar, as permitted by your instructor) for *significant* main effects. In most cases it is not necessary to calculate LSD for judges, and it is not necessary to calculate judge means or judge interaction means. Why?
   (f) Create the appropriate means tables.
   (g) Give these tables numbers and title.
   (h) Arrange means in increasing or decreasing or any logical order, and indicate by underlining or by superscripts which means differ significantly. If using letters or superscipts, means not sharing a common letter are significantly different. You can state this in a footnote.
Examples of statistical tables (Tables 1.1, 1.2, 1.3):

Table 1.2 Means[1] for perceived astringency of Cabernet Sauvignon wines

Wines	CAN winery	HGH winery	LFB winery
Creaminess means[2]	4.3[a]	6.9[b]	8.5[c]

[1]Means sharing superscripts do not differ at $P < 0.05$
[2]Least significant difference = 0.5 at $P < 0.05$

Table 1.3 Means[1] and least significant difference (LSD) for perceived astringency ratings of different Cabernet Sauvignon wines

Creaminess	Mean values
Blue Bell	6.9[b]
Central Dairy	8.5[c]
Hershey's	4.3[a]
LSD	0.5[1]

[1]Means sharing superscripts do not differ at $P < 0.05$

1.5   Graphing Data

Graphs or charts are excellent ways to illustrate information in reports, journal articles, posters, etc. Graphs assist in data interpretation by illustrating trends and relationships among variables and samples. A good graph reveals its message briefly, simply, accurately, and attractively, to improve comprehension of the results. Graphs should never duplicate values shown in a table, and vice versa.

1.5.1 General Rules and Guidelines

The following sequence is important in the development of a good graph:

(a) Determine the significant message you wish to convey.
(b) Determine which type of chart or graph should be used. You should be familiar with available chart types.
(c) Determine who will be the audience and meet that audience on its own level.
(d) Make sure that the data used to make the graph are correct.
(e) Be neat.
(f) Critically evaluate the completed chart, recognize effective results, and strive for improvement.

Consider the following aspects of each graph or chart. Do not accept default options from your graphing program:

(a) Title or caption: Describes the graph simply but completely. It should be possible to read only the title/caption, the axes, and the key or legend to completely understand the graph. Some instructors may require a caption in place of a title. Check with your course syllabus or website.
(b) Axes: Label all axes clearly and correctly.
   (i) X (abscissa) = horizontal. This is usually the independent variable, i.e., the ingredient, process, or time variable that you manipulated
   (ii) Y (ordinate) = vertical. This is usually the dependent variable, which is what you measured, your data, or its summary (such as a mean value). This should go on the left, even if the zero point is in the middle of the graph.
(c) Indicate the zero (0) point, except with logs, years, or category scales for which the lowest value is nonzero (e.g., 1–9). "Break" the scale if necessary.
(d) Show mean values or other measures of central tendency, unless you are making a scatter plot of raw data. Indicate a measure of variability, i.e., standard deviation, standard error of the mean, LSD, etc. The standard error is often a good choice for I-beam error bars

with sensory data because if they do not overlap, then the two points are probably different by a t-test. This permits a quick check of significance "by eye." If instructed to do so, indicate "N," the number of observations that went into calculating the values on the graph.

(e) Use good proportions by choosing the X and Y axes so that the data fill the graph. The ancient Greek proportions of the golden rectangle (approx. 2:3) are appealing to the eye of readers from most Western cultures. Avoid graphs that are too short and flat. Trends are harder to see.
(f) Generally, connected points are better than a smooth curve, although there are exceptions. One exception is when you fit a trend line on purpose, particularly one that is nonlinear or otherwise unexpected.
(g) Usually, place no more than five sets of data per graph. Use different lines or symbols to identify each data set. Give a key or legend to identify the different lines and symbols.
(h) Use log or semilog axes in specialized cases, as appropriate.

1.5.2 Common Types of Charts

1. Line charts: Use with continuous, scaled, or cumulative data to show trends and compare series.
2. Bar charts/histograms: These are column graphs with discrete increments. The independent variables are not necessarily on a continuous scale, and the heights of the bar or dependent measures are often frequency counts. Bars should all be of the same width. Usually, bar graphs show absolute quantities and consist of separate bars, while histograms are used to show percentages or proportions. They may have no spacing between the columns if there is no break in the series. Floating bar graphs have bars that extend above and below a central point.
3. Radar graphs (spider webs): Originate about a central zero point, radiating out in proportion to quantities measured, e.g., polar coordinates.

These are frequently used to report results from descriptive analysis. Circular graphs are usually divided into equal sections. Projections radiate from a central point of origin and distance to the symbols indicates the mean values on that variable or attribute. Specific instructions for graphing spider web plots (also known as radar plots and cobweb plots) using Excel can be found in the Group Project, Descriptive Analysis section.

1.6   Common Problems in Laboratory Reports

1.6.1   Content

1. Failure to do a summary if required.
2. Failure to do a concise literature review when required for the specific report.
3. Failure to include all data analyses and figures. Check with your instructor or course website to determine if raw data tables should be included or appended (or not).
4. Failure to properly label tables, figures, and axes.
5. Failure to critically examine individual as well as group data.
6. Failure to recognize the objectives and principles of the demonstrations.
7. Treating statistical analyses as an end unto themselves, instead of using the statistical results as a tool to help in the interpretation of the experimental results.

1.6.2   Written Text

1. Poor writing ability, leading to long, laborious, and/or redundant explanations.
2. Overuse of personal pronouns, e.g., our data, we found, and my lab section. Substitute: the data shows … the class responded
3. Failure to use the past tense for methods and results (they happened in the past): data were collected, results indicated, analyses were performed, graphs showed, and significance was obtained.
4. Misuse of "between" (two items) and "among" (more than two items).
5. Misuse of "affect" and "effect" and "palate" and "palette" and other homonyms.
6. Misuse of the word data, which is plural. Correct: "The data indicate that…" A datum is singular, Latin, and neuter gender. Data are more than one datum.
7. Spelling, particularly plurals, contractions, possessives and homonyms (to, too, two, do, due, dew). Please use a dictionary and spell-checkers. In addition, you must proofread the report before submission.

1.6.3   Organization and Format

1. Failure to number pages and to number and to appropriately title figures or tables.
2. Careless assembly: sloppy text, blurred photocopies, poor graphs, tattered paper, no staples, no proofreading, etc.

1.7   Appendix: Sample Report (Industrial Style)

Clinton to Bush
12/25/88

Comparison of Discrimination Methods

From: W. J. Clinton, Sensory Department Head
To: G. W. Bush, R & D Manager

Date: December 25, 1988

Background

    Due to increases in sugar prices, cost reductions involving formula changes are of interest. This study examined whether or not a 10% reduction in the sugar content of our Purplestuff drink mix would be perceivably different in taste. A secondary question was the relative sensitivity of several difference testing methods.

Conclusions

1. The lower level of sugar was perceivably less sweet than the current formulation.

2. Paired comparisons were more sensitive than triangle tests, duo-trio tests and rated degree of difference tests.

Recommendations

1. The reduced-level formula should be tested for changes in consumer acceptance.

2. In the future, paired comparisons tests should be performed in situations in which the attribute that is changing is known, and in situations where test sensitivity needs to be maximized.

Full results and procedural details are given in the report appendix.

Clinton to Bush
12/25/88

Appendix

Samples

20 mL samples; room temperature (22 °C), with either 9 or 10% added sucrose (w/v) in Purplestuff drink mix, lot # 123456.

Test Methods

	Null hypothesis	Alt. hypothesis
Triangle	$p = 1/3$	$p > 1/3$
Duo-trio	$p = 1/2$	$p > 1/2$
Paired comparison	$p = 1/2$	$p > 1/2$
Rated difference	9% = 10%	9% ≠ 10%

Results

Triangle	17/44 correct (*NSD*)
Duo-trio	25/44 correct (*NSD*)
Paired comparison	32/44 correct ($p < 0.01$)
Rated difference	Mean (control vs. self) = 1.5
	Mean (sample vs. control) = 1.7
	$t(43) = 0.8$ (*NSD*)

Panelists
44 employee volunteers from the general taste testing pool (untrained).

Lighting
ASTM standard Northern Daylight.

Date of work request: Dec. 23, 1988
Date conducted: Dec. 24, 1988.

Introduction for Instructors and Teaching Assistants

2

2.1 General Considerations and Issues

Various options are possible for lab reports, group work vs. individual work, and reporting. The formats and options given in Chap. 1 illustrate some of the formats that have worked well. It is possible to modify or combine various aspects of the two primary report options. Whether to accept electronic submission is an important issue. Students often prefer sending reports as e-mail attachments, but this may require the TAs or instructor to print out the materials, unless you prefer to look at your computer terminal when grading. Paper submission also allows for handwritten sections, such as calculations or graphs.

The course syllabus, website, or documentation should spell out clearly what report format is required for each lab, as well as expectations for due dates, acceptable modes of document delivery, whether late work will be accepted, any penalties for late submissions, and so on. Another important issue is whether to accept second versions. Labs that are redone can help insure comprehension of important concepts among students who made serious mistakes on the first attempt. One philosophy of education would encourage this although it is obviously more work for the grader. Another option is to accept resubmissions but with an upper limit of 80–90 % of the original grade or point total.

2.2 Specific Options

Several options are possible in conducting these exercises in a sensory evaluation course. The practices you adopt may depend upon your facilities, resources, laboratory personnel, and teaching assistants. Here are several important issues to consider that should be decided well before the semester begins:

1. How much of the analysis are the students responsible for, and how much is to be provided to them? Statistics are often a prerequisite for a sensory course, but if not used often enough, they are forgotten. The lab exercises provide an avenue for reinforcing statistical techniques. On the other hand, if the lab section is large, the data sets may also be large. Then the students may spend a burdensome amount of time number crunching, which can have limited value.
2. What statistical programs will be used, if any? One option is to have all calculations done "by hand," i.e., with a hand calculator. This can teach students how the statistical tests actually work in terms of calculations but can be time-consuming. A second option is to allow spreadsheet programs such as Excel, in order to facilitate taking sums, squaring quantities, etc. but having them show their calculations in the statistical formulae. A third option is to have all students use one statistical package

H.T. Lawless, *Laboratory Exercises for Sensory Evaluation*, Food Science Text Series 2, DOI 10.1007/978-1-4614-5713-8_2, © Springer Science+Business Media New York 2013

and providing them with complete instructions. This approach teaches a certain amount of button pushing, but you can be sure they push the correct buttons. The last approach is to allow students to use any programs they want. Often they may have learned to use a program in a previous statistics class. The disadvantage of this approach is that you may not be able to fathom from whence their mistakes arise, and thus there is no correction learned when they do make mistakes.

A similar line of questions should be asked and answered about graphs and charts that are required. Some of the spreadsheet programs produce graphs that are not always acceptable for scientific publication, at least if one accepts the default settings. Once again, it is up to the instructor to decide how much time he or she wants to spend teaching a specific computer program vs. more basic principles.

3. What preparation and serving responsibilities will be left to the students, and what tasks are to be performed by you, your lab personnel, or teaching assistants? It is not generally practical to have all students make up their own samples for a class. One option is to have them form teams and have each group rotate through responsibilities for some of the kitchen/prep work each week. You must have a certain amount of faith that they will prepare the samples, concentrations, etc. correctly, so some instructors prefer to have teaching assistants or lab staff do the critical work. Of course, it is possible to have students randomize orders and set up trays and tasks that are doable during the class time itself. A common approach is to have students work in pairs, in which one person acts as the server (or experimenter in the old psychophysical jargon) and the other as the panelist or subject. The approach you take may be largely dictated by the resources (lab personnel, teaching assistants) you have at your disposal.

4. Will your lab reports follow one standard format, or does the format depend upon the individual exercise? The advantage to a

standard format (e.g., objective, methods, results, discussion) is that it is easier for students to follow and for graders to grade. But not all exercises fit that mold so well. Some have more statistical analysis; some have less. In the case of the MRE consumer questionnaire exercise, the questionnaire itself is the goal. Section 15.1.1 (the "spam" lab) is only concerned with the development and specification of the test procedure.

One attractive option is to use the labs to teach different writing styles and formats. For example, some instructors require at least one lab in the "industrial report format" which is partially inverted from the scientific journal sequence (conclusions, recommendations, and then all the details). Another useful format is the executive (one-page!) summary, which students often find very difficult, having been rewarded over the years for the length of their reports and papers. Several of these formats are shown in Chap. 1.

2.3 Resources and Requirements

2.3.1 Facilities

Some of these labs are appropriate for a sensory booth-type laboratory room. Others are not. Be careful when choosing a facility or classroom so that the laboratory exercise can be performed in the intended manner. For example, the flavor profile lab requires students to sit around a table to taste the training samples and perform their consensus procedure. It is nearly impossible to do this in a lab with fixed benches or a classroom with fixed chairs as in an auditorium or amphitheater. Obviously, tasting booths are not appropriate either. It is an advantage with these exercises to have some flexibility in the room arrangement or to be able to move the class to the appropriate facility or classroom for each specific exercise. We recognize that this is not always possible in all institutions. Modification of the procedures may be required to fit your resources.

2.3.2   General Supplies and Equipment

The following items are generally useful:

1. A good supply of plastic (recyclable) odorless cups with lids in various sizes such as one-, three-, and six-ounce sizes for serving liquid and semisolid samples. Cups for expectoration should be larger and opaque. The 12-ounce kind that is used for fountain drinks is suitable. It is helpful to have different size rinse and spit cups to prevent students from inadvertently confusing the two. Having someone try to rinse from his or her spit cup is often comic but avoidable.
2. Trays for serving. I prefer white hard plastic dissecting trays. The higher rim on dissecting trays, as opposed to food service trays, can help control spillage.
3. A generous supply of paper goods that would be associated with any test kitchen is useful, as are items to help with spill control. Assume spills will happen.
4. Glassware associated with a chemistry lab is a virtual necessity. 500-ml and 1-, 2-, and 4-l volumetric flasks are useful depending upon the size of your class and the volumes required. Be sure that they are labeled "for food use only" and/or placed in a separate storage area away from chemical/analytical usage. Large beakers, graduated cylinders, and a selection of pipettes of different sizes are also useful.
5. Equipment for weighing samples of various sizes is also needed. Smaller electronic balances with a capacity up to 1 kg are a good idea.
6. Methods to stir, mix, and heat samples to dissolve substances such as sugar are required. Magnetic heated stir plates, stir bars, and stir bar retrievers are very useful.

2.4   Suggested Kitchen/Lab Rules

For kitchen help and prep workers, general food service standards should be adhered to, including hairnets and plastic gloves for prep and serving.

Care should be taken to keep the balances, stir plates, hot plates, and all kitchen equipment clean. Students, including some teaching assistants, will not assume responsibility for earlier spills, often choosing to ignore them, thus leading to accumulated burnt-on grunge. The same is true of microwaves, ovens, and stove tops/ranges. It is useful to establish a kitchen policy as follows: If you find somebody else's mess, clean it up anyway. It does not matter that it is not yours. This is a version of the Boy Scout principle "always leave the campsite cleaner than you found it." People should be discouraged from preparing their lunches in the test kitchen or lab area. It is a common problem to find dirty dishes left (forgotten?) in the kitchen sink, for example. Many students are accustomed to dining halls and/or mommy cleaning up after them. They may assume that dirty dishes will magically disappear, especially if there are full-time staff and/or a lab manager around. Identifying who left the mess may not be possible, so the rule about "if it's dirty you have to clean it even if it's not yours" is applicable here as well. If an automatic dishwasher appliance is present in the kitchen, rules for its use should be clearly posted. When is it to be filled, run, and emptied? What items may or may not be rinsed down the sink should also be posted to prevent clogged drains. Responsibilities should be communicated verbally and reinforced in writing.

2.5   Challenges and Opportunities

The execution of a group project (see Chap. 14) offers many challenges. A common situation occurs when one student fails to contribute his or her fair share to the group effort. I have always viewed this as a valuable life experience, especially for food scientists who will probably end up working on cross-functional teams. One option to penalize the nonparticipating student is to have the group grade each others' efforts, anonymously of course. I have rarely found this necessary although some instructors say it takes care of both the tendency to slack off as well as a general sense of fairness about the project. At the very least, students will learn the value of division of labor.

Whether to ask students to perform classroom presentations is another consideration. Some students find this terrifying, especially those with limited English language skills. Many others, especially seniors, already have experience in slide or PowerPoint presentations and find it less of a burden. As presentation skills (what they used to call "briefing" in the military) are an important part of the skill set for any sensory job in industry, I think the class presentations are a worthwhile conclusion to the group projects.

2.6 Conclusion

Please pay careful attention to the sections marked "notes and keys to successful execution." These notes can help prevent mistakes and situations in which the exercise does not work well or produces useless data. In 20 years my assistants have uncovered most of the possible mistakes. Note that in the list of supplies or materials to shop for, the amounts may need to be adjusted based upon your class size. We assume that most teaching assistants are capable of doing the math. Feel free to modify the exercises, expand, or omit sections as may fit your situation, resources, and the capabilities of your student body.

Part II

Eleven Laboratory Exercises in Sensory Evaluation

Screening Panelists Using Simple Sensory Tests

3

3.1 Instructions for Students

3.1.1 Objectives

To examine methods for screening potential panelists.

To examine tests of odor identification with and without verbal cues.

To examine ranking tests for taste intensity as screening methods.

3.1.2 Background

Panelist selection is an important part of sensory testing. For consumer testing, potential respondents must be screened as users of the product category or perhaps of a specific brand to be in the appropriate reference population for testing. For quality control panels, discrimination testing and for descriptive panels, potential panelists should be physically qualified (i.e., no medical restrictions or allergies associated with the product), and be available, especially if they are employees. For highly trained panels, motivational qualification is also important; sensory testing is hard work that demands concentration and can sometimes be repetitive and tiring.

In addition to the above qualification tests, a basic level of *sensory acuity* is also required of panelists. For discrimination tests, potential panelists should be tested to be sure that their senses are functioning normally. Through screening exercises, we can also get an idea whether or not potential panelists can follow instructions and understand the terminology used. In descriptive analysis or quality control work, the highest performing sub-group of a larger candidate group is often chosen for the panel for subsequent training.

In current practice, it is common to use samples from the actual product category that the panelists will be evaluating, as opposed to using model systems like basic tastes in water. The exercises that follow will use two different screening tests: (1) odor identification tests for determining smell acuity; and (2) ranking tests to determine whether or not panelists can discriminate taste intensities.

Odor identification is a basic skill in sensory work. It is more difficult than most people imagine (Cain, 1979). In everyday life, contextual cues are often available when identifying tastes and odors. When these cues are removed, many people are only able to name about half of the odors they are presented with. However, when a multiple-choice test is given, performance rises to about 75 % correct identifications. If common everyday household odors are used, some people do even better. The enhanced ability seen with a name-matching task shows us that in general, it is difficult to connect odors to verbal processes. They simply use different parts of the brain.

Ranking of sensory intensities is another basic skill. Discrimination between intensities of specific attributes (e.g., sweetness levels) may be required of descriptive panels and in quality control work. Descriptive panels are often trained to use scales with reference to intensity levels of specific standards. Since prospective panelists have not yet been trained, they should not be expected to use a scale; but correct rankings of intensity levels is a reasonable task to put to the candidates.

3.1.3 Materials and Procedures, Part 1: Odor Identification

3.1.3.1 Materials (For Each Group of 5–6 Students)

Sets of six jars with lids containing unique odors (all labeled "A," individually labeled with unique three-digit codes).

Second sets of six jars with lids containing unique odors (all labeled "B," individually labeled with unique three-digit codes).

Ballots consisting of two blank pages labeled A and B.

3.1.3.2 Procedures

Begin with odor set A. Smell the odors in each of the screw cap jars marked A. Try to identify each odor by writing a word or two on the ballot that best describes each odor. Be sure that the three-digit code on the lid and the jar matches the code for the answer line on the ballot. Form groups to share a set of bottles by passing them around in a circle, but do not discuss your impressions with your neighbors, please. Next, using odor set B, smell the odors in each of the screw cap jars. Try to identify each order from the list of selections provided as hints. Consider this a multiple choice test. When you have completed both exercises, we will discuss the correct answers, and then you will give your ballot to the instructor or to a TA for tabulating.

3.1.4 Materials and Procedures, Part 2: Taste Ranking

3.1.4.1 Materials (For Each Individual)

Three samples of apple (or other fruit) juice: One as purchased, one with 0.5 % added sucrose and one with 1.0 % added sucrose (wt/vol).

Three samples of apple (or other fruit) juice: one as purchased, one with 0.1 % added tartaric acid and one with 0.2 % added tartaric acid.

3.1.4.2 Procedures

Rank the first set of three apple juices for sweetness per the instructions on the ballot (3 = most sweet; 1 = least sweet).

Rank the second set of three apple juices for sourness per the instructions on the ballot (3 = most sour; 1 = least sour)

3.1.5 Data Analysis

Obtain frequency counts of the numbers correct from your instructor or website.

Taking the data from each person as a pair of observations, perform a paired t-test on the data from the unaided and multiple choice versions of the odor identification. Calculate the correlation coefficient between the two conditions using the same data. For the taste ranking, you will construct a simple table of the number of correct rankings by the "correctness" as determined by the number of reversals, as illustrated below.

3.1.6 Reporting

Use the standard lab report format unless instructed otherwise. Provide a brief discussion in complete English sentences to accompany your graphs and tables. Answer the questions raised below in your results and discussion sections.

1. Odor identification results:
   (a) Plot a histogram is a bar graph that shows a frequency distribution of the data. Plot the number of correct odor identifications (zero to six) from each set on the x-axis and the number of people that got each of those scores on the y-axis (frequency). Construct separate histograms for the free-choice and the matching portions of the exercise. You can use Excel or some other graphing program to do this or you can draw them by hand. If hand drawn, use a straightedge for axis lines and any borders.

Hand-drawn graphs should be neat, with ruler-guided lines and legible axis labels.

(b) Perform a paired *t*-test using (3.1) (for additional help, see Statistical Appendix A of Lawless and Heymann 2010) to compare the mean number of correct odor identifications in the free-choice procedure with the mean number of correct odor identifications in the matching procedure. Was mean performance significantly higher for one or the other method?

(c) Calculate the correlation coefficient (see Statistical Appendix D of Lawless and Heymann 2010) between the two odor identification methods. Did people who performed well in one method also perform well in the other method? Would you combine individual performances in the two methods to achieve an individual's total score? Why or why not? (hint: if they are correlated they may be tapping into the same ability).

2. Taste ranking results

(a) Check the scoring key to see if your rankings were correct or incorrect. It is possible to have perfect rankings (e.g. 123), one reversal (e.g. 132 or 213), two reversals (e.g. 231 or 312) or complete (three) reversals (321). The TA will tabulate the numbers correct for sweetness ranking, sourness ranking and class totals for both. You will need this information to answer the following questions. Make a simple table of the number correct vs. number of students who scored that way for sweetness and sourness rankings.

(b) Answer the following questions:
The chance of getting the ranking perfectly correct is one in six by guessing. Did the class do better than one-out-of-six correct? Would you use this test in screening for apple juice judges in a QC operation? (Why or why not?) How could you modify the test to make it better?

Useful equations:

$$t = \frac{\bar{D}}{\frac{\sigma_{diff}}{\sqrt{N}}} \quad (3.1)$$

Where t has $N = 1$ degree of freedom for N pairs, $\bar{D}$ is the mean of the difference scores and σ_{diff} is the standard deviation of the difference scores.

$$r = \frac{\sum XY - \left(\frac{\sum X \sum y}{N}\right)}{\sqrt{\left[\sum X^2 - \frac{(\sum X)^2}{N}\right]\left[\sum Y^2 - \frac{(\sum Y)^2}{N}\right]}} \quad (3.2)$$

3.2 For Further Reading

Cain WS (1979) To know with the nose: keys to odor identification. Science 203:467–470

Lawless HT, Engen T (1977) Associations to odors: Interference, mnemonics and verbal labeling. J Exp Psychol Human Learn Mem 3:52–57

Lawless HT, Heymann H (2010) Sensory evaluation of foods, principles and practices, 2nd ed., Springer Science+Business, New York

3.3 For Instructors and Teaching Assistants

3.3.1 Notes and Keys to Successful Execution

1. The percent correct in the un-cued odor ID condition is generally 50–75 % (four out of six is common) but somewhat higher when the verbal cues are available. The handout or overhead list of cues for set B must of course include all of the odors in Set B. Some of the students will experience the "tip of the nose effect" in which they seem to know they are familiar with the odor but can't find the name (Lawless and Engen, 1977). This is an opportunity for discussing the olfactory-verbal gap and the difficulty in naming odors, as well as the need for training for odor and flavor description.

2. There is a suggested list of odors for Set B in the ballots and data sheets appendix.

3. Be sure that students understand what a frequency histogram is. If they simply put the

raw data into Excel, it will not make the correct graph.

4. Do not allow prep personnel to add sucrose to the desired (i.e. final) amount of apple juice. The sugar will take up space and increase the final volume, so the concentration will not be accurate. The correct method is to take a smaller volume (say 75 %) of the desired final amount, add the sucrose while stirring, and then top it off to get to the final desired volume. Larger size volumetric flasks are recommended (2–4 L).

3.3.2   Equipment

For Part 1, odor identification, no special equipment is required.

For Part 2, volumetric flasks (1 L or larger), stir plates, stir bars, storage containers, label gun or other means of labeling cups with random codes. Sample trays for each student are recommended.

3.3.3   Supplies

Part 1

Perfumer's paper blotting strips or unscented Q-tips or cotton balls.

Sets of 6-oz amber jars with screw caps, preferably Teflon lined (12 per groups of 5–6 students).

12 liquid odorants, flavor extracts or similar familiar-smelling liquids.

Part 2

Apple juice, about 5–6 L for class of 25; Sucrose, commercial grade is acceptable; Tartaric Acid, food grade. Sample cups (30 ml or larger), rinse cups, spit cups, water, napkins, crackers, spill control and garbage receptacles.

3.3.4   Procedure

Part 1

Place a drop or two of each odorant on the perfumer's blotter or cotton ball and place each one in a uniquely coded amber jar. Each set of 6 should be diverse and each set should be of about the same difficulty level. Try to avoid odors that some people may not be familiar with such as lavender. All of the items in Set B should be on an overhead or handout. Do not show the overhead or give the handout until Set A (the un-cued set), has been completed. If the choices are printed on the ballot, DO NOT give out the Set B ballot until Set A has been completed.

Part 2

Make solutions by taking about 75 % of the desired final volume, add the sucrose or acid while stirring, and then top off (still stirring) to get the final desired volume. A volumetric flask is recommended. You cannot simply add the sucrose to the desired final volume because the liquid will expand as the sugar is added and the concentration will not be accurate.

5 l of apple juice will be enough to make 45 cups of each of the samples below (each cup containing 20 ml sample) (Table 3.1).

Table 3.1 Suggested codes for taste ranking test samples

Code number	Contents
582	Control sample (as purchased)
683	1 % added sucrose (apple juice + 10 g sucrose per liter of final product)
815	2 % added sucrose (apple juice + 20 g sucrose per liter of final product)
869	Control sample (as purchased)
673	0.1 % added tartaric acid (apple juice + 1 g tartaric acid per liter of final product)
174	0.2 % added tartaric acid

3.4   Appendix: Sample Ballots and Data Sheets

Response Sheet   *Name (or ID):*

Odor Identification (SET A)

Smell the samples in Set A and identify the perceived odor. Write your answers in Table 1.

Table 1. Free-Choice Procedure (Set A)

Sample Code	Perceived Odor
163	
825	
287	
907	
653	
197	

Response Sheet   *Name (or ID):*

Odor Identification (SET B)

Smell the samples in Set B and identify the perceived odor by <u>matching</u> to the list provided on the overhead or handout. Write your answers in Table 2.

Table 2. Matching Procedure (Set B)

Sample Code	Perceived Odor
479	
509	
688	
109	
621	
774	

Response Sheet   *Name:*

Taste Ranking

Rank the samples in terms of most to least **SWEET**. Write the sample numbers in the space provided.

    Most sweet   Least sweet

    _____   _____   _____

Rank the samples in terms of most to least **SOUR**. Write the sample numbers in the space provided.

    Most sour   Least sour

    _____   _____   _____

(Blank) *Answer Keys*

Odor Identification
Table 1. Free-Choice Procedure (Set A)

Sample Code	Substance
163	
825	
287	
907	
653	
197	

Table 2. Matching Procedure (Set B)

Sample Code	Substance
479	
509	
688	
109	
621	
774	

Taste Ranking (fill in three digit codes)

SWEET
    Most sweet   Least sweet

    _____   _____   _____

SOUR
    Most sour   Least sour

    _____   _____   _____

Student Data Sheet

Odor Identification

No.	Name or ID #	Number Correct (out of 6)	
		Free-Choice--Set A	Matching--Set B
1			
2			
3			
4			
5			
6			
7			
8			
9			
10			
11			
12			
13			
14			
15			
16			
17			
18			
19			
20			
21			
22			
23			
24			
25			
26			
27			
28			
29			
30			
31			
32			
33			
34			
35			
36			
37			
38			
39			
40			

Student Data Sheet

Taste Ranking

No.	Name or ID #	Reversals		
		SWEET	SOUR	Total
1				
2				
3				
4				
5				
6				
7				
8				
9				
10				
11				
12				
13				
14				
15				
16				
17				
18				
19				
20				
21				
22				
23				
24				
25				
26				
27				
28				
29				
30				
31				
32				
33				
34				
35				
36				
37				
38				
39				
40				

Comparison of Discrimination Test Methods

4

4.1 Instructions for Students

4.1.1 Objectives

To become familiar with four different discrimination test methods.
To compare the relative sensitivities of four different discrimination test methods.
To prepare a report of lab results in an industrial report format.

4.1.2 Background

Discrimination tests are commonly used to determine whether or not small changes in ingredients and/or processing or packaging techniques have any effect on the sensory characteristics of a particular product. A manufacturer may wish to make any number of these changes as suppliers or ingredients change, to produce a more nutritionally beneficial versions of the product or to reduce costs.

There are several ways to test for perceivable differences and to make the test objective rather than subjective (Peryam and Swartz, 1950). In one kind of test, a respondent is required to choose a sample from a group of two or more samples that has more or less of a particular attribute. This is called an *n*-AFC test for *n*-alternative forced choice (*n* being usually from 2 to 4, e.g., 2-AFC). In another case, the respondent chooses a sample that is different somehow from the other samples in the group. An example is the triangle test in which two samples are from the same treatment or batch and the third sample is a different product. In a third kind of test, there is a matching requirement: the respondent must match one or more of the test items to one or more reference items that were previously tasted. The latter type includes duo-trio, ABX, and dual standard tests.

All of these tests have chance performance levels, that is, the proportion correct that would be expected if there were no difference and people were forced to guess. Thus, a null hypothesis states that the population proportion correct is expected to be the chance performance level, for example, 1/3 in the triangle test. Note that this null is not a verbal statement ("no difference") but a specific mathematical relationship, an equation. The tests are generally one-tailed, meaning the alternative hypothesis is that the population proportion correct is *greater than* the chance level (and not "not equal to"). So the alternative hypothesis is a mathematical inequality.

Not all discrimination test methods perform equally well in detecting small differences between products (Ennis, 1993). Some methods, such as the triangle test, require difficult comparisons by the respondent. In theory, three pairs must be considered in the triangle test to determine which item is the outlier and which pair contains the duplicates. Other test methods, such as the paired comparison, merely require respondents to "skim" the sensations to determine which was strongest or weakest. Therefore, given equivalent chance performance levels, one test may still be more sensitive to differences than another

(O'Mahony and Rousseau, 2002). The paired comparison test, for example, tends to be more sensitive to differences than the duo-trio test, even though they both have the same chance performance level (i.e., $p = 0.5$).

In the exercise that follows, a test and control product will be used in four different discrimination test procedures. The difference between the test and control products involves a small ingredient change.

The discrimination tests include the triangle test, the dual standard test, a 3-AFC test, and a 2-AFC or paired comparison test. Consult your instructor or lab website in case these have changed from previous years. The tests will be performed independently by each student in the order listed above. Additional background information about discrimination test methods is found in Lawless and Heymann (2010), Chaps. 4 and 5.

4.1.3 Materials and Procedures

4.1.3.1 Materials

Obtain from a TA the following materials:

A white tray containing 11 samples of a powdered fruit drink (*Before you begin* make sure all codes on your tray correspond with codes on the 4 ballots you receive.)

Four ballots (one for each test) each with instructions on the appropriate testing method

Water, crackers, napkins, and spit cups

4.1.3.2 Procedures

Perform the tests as instructed on the individual ballots in the following order:

Triangle test, dual standard, 3-AFC test, and paired comparison test.

Once you have completed the four test methods, give your ballots to a TA. The TAs will tabulate the numbers correct in each test and will send you the results by e-mail or post them on your course website.

4.1.4 Data Analysis

Determine the percent correct responses for each test method. Place in a table, along with the number of respondents, n, and the following information.

Determine whether or not each test method found the two products to be statistically significantly different from one another. Use the following binomial approximation to the normal distribution, with a critical z-value of 1.645 for significance:

$$z = \frac{(P_{obs} - P_{chance}) - (1/2n)}{\sqrt{\dfrac{pq}{n}}}$$

where P_{obs} is the proportion correct you found for each test, P_{chance} is the chance probability for each test ($= p$ in the denominator), $q = 1 - p$, and n is the number of judges.

Check your results against the tables of "minimum numbers of correct judgments" found in Lawless and Heymann to make these determinations.

Calculate the *estimated proportion of discriminators, D/n*, for each method given the obtained results (use the formula below):

$$C = D + p(n - D)$$

where C = the number of correct responses, D = the number of discriminators, n = the number of respondents, p = the chance performance level of the test (1/2 or 1/3)

Note: If the estimated number of discriminators is less than zero, report the proportion as zero.

4.1.5 Reporting

Put the first three analyses in the Results section of your report in your table. The recommended report format for this report is the industrial report format. However, you may use the standard report format if your class website or syllabus instructs otherwise.

A two-page industrial report based on the above experimental procedures and results should include the following sections:

Author/Title/Date. Also, to whom is this addressed?

Background (you may invent a short scenario here)

Conclusions

Recommendations

Methods

Results

Literature (references—who or what should you cite?)

A sample industrial report is seen in the introduction Chap. 1 appendix.

Be careful! Your report may not match this one in terms of specific items, so look over your report and make sure each part is actually what you did.

1. Be sure to limit your industrial report to two pages. Extra pages are not acceptable.
2. Be sure to answer the following questions:
   (a) Was there evidence of a difference between the two products?
   (b) Which test method appears to be the most sensitive? Why do you think that this is the case? Put this answer in the Conclusions section of your report.

4.2 For Further Reading

Ennis DM (1993) The power of sensory discrimination methods. J Sens Stud 8:353–370

Lawless HT, Heymann H (2010) Sensory evaluation of foods, principles and practices, 2nd ed., Springer Science+Business, New York

O'Mahony M, Rousseau B (2002) Discrimination testing: a few ideas, old and new. Food Qual Prefer 14:157–164

Peryam DR, Swartz VW (1950) Measurement of sensory differences. Food Technol 4:390–395

4.3 For Instructors and Assistants

4.3.1 Notes and Keys to Successful Execution

1. Any three-digit random codes may be substituted. If needed, the codes can be changed from year to year to discourage blatant copying of previous lab reports. You can also change the tests. One option is to swap out the dual standard for the ABX test, which is very similar. The triangle and 3-AFC should be included due to the Gridgeman's paradox result (3-AFC better than triangle results) and the opportunity to discuss the said paradox.

2. The volumes given below are minimum amounts for a class size of 35. Adjust as needed based upon the class size. Remember the size of a typical male sip is 25 ml and the female sip about 15 ml.

3. It is important to start with *unsweetened* Kool-Aid or some other powdered drink mix. Beware of products labeled simply "sugar-free" as they may already be sweetened with aspartame or some other intensive sweetener. Using such a product is a fatal mistake!

4. Sucrose percents are weight per volume, for example, 10 % sucrose is 10 g of sucrose in 100 ml *final volume* (100 g/L, final volume). Do not add 10 g to 100 ml, but start with 10 g and mix slowly adding water to get 100 ml at the end. The water will expand as sugar is added due to the partial molar volume of the sucrose.

4.3.2 Equipment

Label gun or other means of affixing 3-digit random code labels to cups.

Trays for each student are recommended to set up samples ahead of time.

Balance for weighing sucrose and volumetric flasks for dissolving.

Hot/stir plate and stir bars.

4.3.3 Supplies

Kool-Aid or any powdered drink mix, unsweetened, any flavor.

Sucrose. Commercial cane sugar is acceptable.

1–3-oz (30–100 ml) plastic cups for test samples.

8-oz (240 ml or larger) plastic cups for rinse water.

Spring water or other odorless water of higher purity.

Paper for ballots.

Napkins, spill control, and cleanup supplies.

Table 4.1 Suggested sample codes and test products

Sample code	Test product
Triangle test (give any three, ask to choose which one is different, see sample ballots)	
469	Powdered drink mix at 9 % sucrose wt/vol
642	Powdered drink mix at 9 % sucrose wt/vol
849	Powdered drink mix at 10 % sucrose wt/vol
703	Powdered drink mix at 10 % sucrose wt/vol
Dual standard test (ask to match to reference, see sample ballots)	
Ref A	Powdered drink mix at 9 % sucrose wt/vol
811	Powdered drink mix at 9 % sucrose wt/vol
Ref B	Powdered drink mix at 10 % sucrose wt/vol
837	Powdered drink mix at 10 % sucrose wt/vol
3-AFC test (ask to choose which one is sweetest, see sample ballot)	
679	Powdered drink mix at 9 % sucrose wt/vol
995	Powdered drink mix at 9 % sucrose wt/vol
685	Powdered drink mix at 10 % sucrose wt/vol
2-AFC test (ask to choose which one is sweetest, see sample ballot)	
824	Powdered drink mix at 9 % sucrose wt/vol
762	Powdered drink mix at 10 % sucrose wt/vol

4.3.4   Sample Preparation

Dissolve two sachets of powdered mix in about 3 L of water and add 360 g sucrose. Stir on a stir plate and add water to achieve a final total volume of 4 L. Final concentration = 9 % wt/vol.

Dissolve two sachets in about 3 L of water and add 400 g sucrose. Stir on a stir plate and add water to achieve a final total volume of 4 L. Final concentration = 10 % wt/vol.

If sachets are not used but another form of container, follow the manufacturer's directions for the desired volumes, except you should add the correct amount of sucrose as indicated above.

Table 4.1 shows the entire list of needed samples and some suggested three-digit codes.

This should be enough to make 35 trays, with each tray containing 11 cups and each cup containing 20-ml sample. You may adjust the volumes necessary for the size of the class.

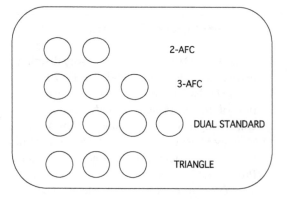

Fig. 4.1 Indicates a tray setup, in which the samples are tasted from *left* to *right* and *front* (*bottom*) to *back* (*top*)

Fig. 4.1 shows a sample tray. A suggested serving order (randomization) is indicated on the sample ballots in the appendix of this chapter.

4.4 Appendix: Ballots and Tabulation Sheets

Triangle Test ✄

Please rinse your mouth with water before beginning. You should have three samples in front of you. Two of these samples are the same and one is different. Please taste the samples in the order presented, from left to right. Circle the number of the sample that is different. Please rinse your mouth with water between samples and eat a cracker if necessary.

469   624   849

✄ -

Triangle Test

Please rinse your mouth with water before beginning. You should have three samples in front of you. Two of these samples are the same and one is different. Please taste the samples in the order presented, from left to right. Circle the number of the sample that is different. Please rinse your mouth with water between samples and eat a cracker if necessary.

703   624   469

✄ -

Triangle Test

Please rinse your mouth with water before beginning. You should have three samples in front of you. Two of these samples are the same and one is different. Please taste the samples in the order presented, from left to right. Circle the number of the sample that is different. Please rinse your mouth with water between samples and eat a cracker if necessary.

469   849   624

Triangle Test

Please rinse your mouth with water before beginning. You should have three samples in front of you.
Two of these samples are the same and one is different. Please taste the samples in the order
presented, from left to right. Circle the number of the sample that is different.
Please rinse your mouth with water between samples and eat a cracker if necessary.

469   703   849

✂ - .

Triangle Test

Please rinse your mouth with water before beginning. You should have three samples in front of you.
Two of these samples are the same and one is different. Please taste the samples in the order
presented, from left to right. Circle the number of the sample that is different.
Please rinse your mouth with water between samples and eat a cracker if necessary.

703   624   849

✂ - .

Triangle Test

Please rinse your mouth with water before beginning. You should have three samples in front of you.
Two of these samples are the same and one is different. Please taste the samples in the order
presented, from left to right. Circle the number of the sample that is different.
Please rinse your mouth with water between samples and eat a cracker if necessary.

849   703   624

Dual Standard ☆

Please rinse your mouth with water before beginning. You should have four samples in front of you.
Two of the samples have a 3-digit code. The other two samples are references "A" and "B", each of
which match one of the coded samples.
First, taste each of the reference samples. Then, taste the coded samples. Write down the correct
reference to the coded sample in the space provided.
Please rinse your mouth with water between samples and eat a cracker if necessary.

_____ matches 811

_____ matches 837

✂ ---

Dual Standard

Please rinse your mouth with water before beginning. You should have four samples in front of you.
Two of the samples have a 3-digit code. The other two samples are references "A" and "B", each of
which match one of the coded samples.
First, taste each of the reference samples. Then, taste the coded samples. Write down the correct
reference to the coded sample in the space provided.
Please rinse your mouth with water between samples and eat a cracker if necessary.

_____ matches 837

_____ matches 811

3-AFC Test ✄

Please rinse your mouth with water before beginning. You should have three samples in front of you.
Please taste the samples in the order presented below from left to right. Within the three samples, circle
the sample that tastes the sweetest.
Please rinse your mouth with water between samples and eat a cracker if necessary.

              685   679   995

✄ - .

3-AFC Test

Please rinse your mouth with water before beginning. You should have three samples in front of you.
Please taste the samples in the order presented below from left to right. Within the three samples, circle
the sample that tastes the sweetest.
Please rinse your mouth with water between samples and eat a cracker if necessary.

              995   685   679

✄ -.

3-AFC Test

Please rinse your mouth with water before beginning. You should have three samples in front of you.
Please taste the samples in the order presented below from left to right. Within the three samples, circle
the sample that tastes the sweetest.
Please rinse your mouth with water between samples and eat a cracker if necessary.

              995   679   685

Paired Comparison Test ✄

Please rinse your mouth with water before beginning. You should have two samples in front of you.
Please taste the samples in the order presented below from left to right. Within the pair, circle the sweeter sample.
Please rinse your mouth with water between samples and eat a cracker if necessary.

824   762

✄ -

Paired Comparison Test

Please rinse your mouth with water before beginning. You should have two samples in front of you.
Please taste the samples in the order presented below from left to right. Within the pair, circle the sweeter sample.
Please rinse your mouth with water between samples and eat a cracker if necessary.

762   824

Data Sheet-

No.	Name or ID number	Please mark each test that you answered CORRECTLY			
		Triangle	Dual Standard	3-AFC	Paired Comparison
1					
2					
3					
4					
5					
6					
7					
8					
9					
10					
11					
12					
13					
14					
15					
16					
17					
18					
19					
20					
21					
22					
23					
24					
25					
26					
27					
28					
29					
30					
31					
32					
33					
34					
35					

5.1 Instructions for Students

5.1.1 Objectives

To become familiar with a rapid method for estimating taste or smell thresholds from a group.

To understand individual differences in chemosensory acuity.

To become familiar with an ASTM procedure.

5.1.2 Background

Thresholds are generally defined as the minimal level of a stimulus that is detected on 50 % of trials by an individual or by 50 % of a group of people. Detection thresholds are useful measuring tools for determining the range of minimum levels of a stimulus that most people would perceive. Specific applications of detection thresholds in sensory science include:

- Specifying the potency or biological activity of a flavor compound in food
- Determining the presence of a taint or off-flavor in a food that has deteriorated
- Specifying the minimum acceptable levels of such taints and off-flavors [see Stocking et al. (2001)]
- Measuring an individual's sensitivity to some flavor compound of interest

People differ in their sensitivities to taste and aroma compounds. A classic example is the bitterness blindness to phenylthiocarbamide (PTC, phenylthiourea) and to propylthiouracil (PROP). Another example is specific anosmia (specific smell blindness) to structurally related families of chemical compounds. Anosmia to short-chain fatty acids, such as isovaleric, is one example. This condition exists when a person or group has a threshold value two standard deviations higher than the population mean.

Different threshold procedures can give different results. Response biases, choice of statistical cutoffs, and details of presentation procedures are a few of the factors that complicate the measurement of sensory thresholds. Several practical methods have been proposed that are discussed at length in Lawless and Heymann (2010), Chap. 6. This lab exercise will use the ASTM E-679-79 procedure, but with two different methods of calculating the resulting threshold value. The American Society for Testing and Materials (now known as ASTM International) is an industrial standards organization that publishes consensus standards for materials and testing methods. Although ASTM started primarily as a standards organization for building materials, they have expanded to numerous products and testing procedures and include a very active group developing consensus sensory methods called Committee E-18. Their methods are found in volume 15.08 of the ASTM books of standards (ASTM, 2008).

This method will attempt to estimate a group threshold from a small panel (i.e., from class data). A recommended panel size for this kind of threshold estimate is about 25 people, unless wide individual differences are expected, as in specific anosmia. There are two threshold estimates you can obtain from these data, one using the ASTM group average of the individual BET (Best Estimate Thresholds) and the second using a plot of the proportion correct at each concentration step and interpolating at the 66.6 % correct level, to adjust for the probability of guessing correctly.

The individual BET is defined as the geometric mean (nth root of the product of n values, i.e., the square root of the product of two values). In this method, we interpolate between two adjacent concentration levels: the first value that had all higher values chosen correctly and the last incorrect step below that string of correct choices. There are also a few "rules of thumb" in this method. If a person misses the highest concentration level, the BET is the geometric mean of the highest step and the next step that would have been used if the series had been continued upward. If a person gets all seven steps correct, the BET value is the geometric mean between the first step and the one step lower that would have been used if the series had been extended downward. In other words, extrapolations are used for perfect performance and for complete inability to distinguish the highest level in the study.

The data set has additional information. For each concentration step, we can also find the group proportion correct; plot this as a function of concentration. Next, we define a proportion as the threshold concentration and interpolate to estimate what concentration would give that level of correct choices. When we use a three alternative test with a 1/3 chance of correct guessing, the appropriate proportion is 66.7 % correct. A chance-corrected level of 66.7 % is based on the correction for guessing needed to obtain 50 % discriminators from a 3-AFC test. See Antinone et al. (1994) and Lawless (2010) for other applications of this rule. The correction for guessing is given by Abbott's formula as follows:

$$P_{required} = P_d + P_{chance}(1 - P_d) \qquad (5.1)$$

where P_d is the proportion of discrimination that we wish to achieve (50 % for thresholds).

In the case of a 3-AFC test, with $P_{chance} = 1/3$, we get $0.5 + 1/3(1 - 0.5) = 2/3$ or 66.7 %.

Table 5.1 Concentration steps for the threshold series

Step number	Log molarity	mg/l
1	−7.0	0.07
2	−6.5	0.21
3	−6.0	0.68
4	−5.5	2.15
5	−5.0	6.79
6	−4.5	21.46
7	−5.0	67.86

5.1.3   Materials and Procedures

5.1.3.1 Materials
Seven concentration steps of sucrose octaacetate (SOA) in springwater as shown in Table 5.1.

5.1.3.2 Procedures
1. Each student should obtain from a TA the following items:
   (a) A tray containing fourteen 10-mL samples of spring water (blanks) and seven 10-mL samples of SOA. (All samples are identified only with random three-digit codes.)
   (b) An ascending method of limits ballot containing the codes on the 21 sample cups organized into seven groups of three.
   (c) A key identifying which sample in each group of three contains SOA. Please do not look at the key until after you have made your judgments.
2. Taste the samples in the order on the ballot (left to right); identify the sample in each group of three that is the most different from the other two. If the student cannot decide which sample in each group of three is the most different, he or she must guess.
3. Rinse your mouth with water and wait about 30 s between sets of three samples. *(All of the above procedures are reiterated on the ballot and on the key)*
4. Repeat the tasting for the next triad, working from the bottom (front) of your set to the top.
5. After recording your choices, refer to the answer key and mark your choice as correct or incorrect at each step and record on the class master data sheet.

5.1.4   Data Analysis

There are two threshold estimates you can obtain from these data, one using the ASTM group average of the individual BET (best estimate thresholds) and the second using a plot of the proportion correct and interpolating at the 66.6 % correct level. Use the class data to perform the following analyses, to be entered in your results section:

1. Plot a histogram of *individual* thresholds. Do they appear to be normally distributed or were there some outliers? Calculate the first group threshold as the geometric mean of the individual BET thresholds posted in the class data.

   Note: the geometric mean can be calculated from taking the arithmetic mean of the BET thresholds expressed as log values, then taking the antilog (10^x of that value).

   For example, the *arithmetic mean of the log values* might be $[-4.25 + -5.75 + \ldots]/n = -4.25$; it then should be expressed in moles/liter as the *antilog of the mean log threshold*, or $10^{-4.25} = 56.23$ µmol/l. Be careful to recognize what units you are using. DO NOT calculate the log of the log values first.

2. Plot a graph of the *group percent correct responses* (*y*-axis) as a function of concentration *level of sucrose octaacetate* (*x*-axis) using a log scale for concentration.

   Pass a "line of best fit" (using graphing software such as MS Excel or "by eye" if you do not have curve-fitting capabilities) through the plot of *group percent correct responses* and interpolate from this line the concentration of SOA to which 67 % of the group would respond correctly. See Fig. 5.1 for an example of the interpolation. The 67 % level is the chance-corrected level for 50 % detection after correcting for guessing (see Chap. 6 in Lawless and Heymann and Lawless (2010) for further examples).

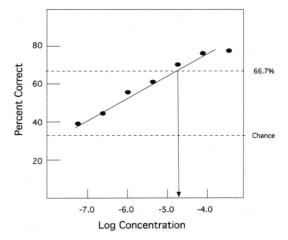

Fig. 5.1 Example of interpolation at the 66.7 % correct level to find the chance-corrected level for 50 % discriminators in a 3-AFC task

your calculations, and answer the two discussion questions in complete English sentences. One page (typed) of discussion is sufficient.

Discussion Questions

Did the two methods of assessing group thresholds agree with one another? Why do you think they did or did not?

Were there any flaws or problems you encountered using these methods to determine thresholds?

Extra Credit

How could you estimate the level of error in the threshold concentration around your two measurements?

Hint: Error should be expressed in concentration units, ±X moles/liter, not as standard deviations around the sensory data. Remember that the data are in percent correct, not a concentration measure. So you must find a way of getting from variability on the *y*-axis to variability expressed on the *x*-axis. Be specific as to your method and/or calculations.

5.1.5   Reporting

Use the standard report format unless your instructor requires otherwise. Include the graphs and analyses required above in your results section. Please turn in your two graphs, show all

5.2   For Further Reading

Antinone MA, Lawless HT, Ledford RA, Johnston M (1994) The importance of diacetyl as a flavor component in full fat cottage cheese. J Food Sci 59:38–42

ASTM (2008) Standard practice for determining odor and taste thresholds by a forced-choice ascending concentration series method of limits, E-679-04, Annual Book of Standards, Vo.15.08, ASTM International, Conshocken, PA, pp. 36–42

Lawless HT (2010) A simple alternative analysis for threshold data determined by ascending forced-choice method of limits. J Sens Stud 25:332–346

Lawless HT, Heymann H (2010) Sensory evaluation of foods, principles and practices, 2nd ed., Springer Science+Business, New York

Stocking AJ, Suffet IH, McGuire MJ, Kavanaugh MC (2001) Implications of an MTBE odor study for setting drinking water standards. J AWWA 2001:95–105

5.3   For Instructors and Assistants

5.3.1   Notes and Keys to Successful Execution

1. A parent or stock solution can be prepared using 10 % ethanol in springwater to first dissolve the SOA. This dilution step was $100X$ the concentration of the highest SOA concentration used in this exercise. It is recommended that serial dilutions be made from a concentrated stock solution or from multiple stock solutions after serial dilutions rather than trying to weigh out minute amounts of SOA on an analytical balance. Sucrose octaacetate is a potent bitter substance with very low (essentially zero) toxicity. If swallowed, it is hydrolyzed in the digestive tract to minute amounts of sucrose and acetic acid.

2. Spitting is recommended in this lab, primarily to prevent buildup or carry-over from bitter tastes in the throat.

3. Note that although the test is set up as if it was a 3-AFC test, with two water blanks in each triad, the ASTM instruction specify choosing the sample "that is most different" from the other two, as if it was a triangle test. Thus, there is no single Thurstonian model for this test, as it is unclear whether the subjects will use a distancing strategy as specified by the instructions or a skimming strategy as suggested by their knowledge that there are two blanks.

4. The line of best fit for the group percent correct responses can be fit by a logistic regression in which $\ln(p/(1-p)) = b_o + b_1 \log C$, where p is the percent correct, b's are slope and intercept, and C is concentration. Solving for the 66 % or 2/3 correct level is simply solving for $\ln(66/33) = \ln(2)$.

5. The reference for Lawless (2010) shows these methods in greater detail and discusses the pros and cons of using the ASTM BET method vs. the chance-corrected interpolation.

6. Error can be estimated by converting back from a confidence interval or SDs around the sensory data to the corresponding concentrations. For the group plot of percent correct, the error around the line can be expressed as a function of the standard error of a proportion (SQRT(pq/n)). This can be used to set up a lower and upper error envelope around the fit line. Then the 66 % level can be used to interpolate an upper and lower concentration from the intercepts with the upper and lower envelope curves.

5.3.2   Equipment

Stir plate (s), spin bars, and weighing balance.
Glassware for serial dilutions. Volumetric flasks are recommended.
Pipettes and/or graduated cylinders.
Trays are recommended for each student's samples.

5.3.3   Materials and Supplies

Spring water or odorless water of greater purity.
Sucrose octaacetate (CAS 126-14-7) and food grade ethanol (95 %).
Cups for rinse water (large) and sample cups (small, to hold 10–20 ml).
Rinse water, spit cups, napkins, and unsalted crackers.
Spill control.

5.3.4   Procedure

See notes for hints about making serial dilutions from a concentrated stock solution. Once the first three or four dilutions have been made, the third or fourth can serve as a stock solution for the lower ones. Make sure that the new solutions are well mixed. Instructors and/or TAs should taste the resulting series to insure that the SOA is detectable at the higher levels.

For each student, set up a tray containing fourteen 10-mL samples of spring water (blanks) and seven 10-mL samples of SOA. (All samples are identified only with random three-digit codes.)

Provide a ballot containing the codes on the 21 sample cups organized into seven groups of three. Organize these in rows so that each student will work from bottom (front) to top (back), choosing the odd sample in each row.

Prepare a key identifying which sample in each group of three contains SOA. Two sample keys are shown in the appendix.

5.4   Appendix

Response Sheet 1

Taste the samples in the order on the ballot, <u>from left to right</u>. If you have a tray, check to see that the Key numbers match the ones on your samples.

Circle the number of sample in <u>each</u> level is the most different from the other two. If unsure you must guess.

Rinse your mouth with water and wait about 30 seconds before moving to the next Level.

LEVEL 1	247	188	633
LEVEL 2	172	962	741
LEVEL 3	657	790	569
LEVEL 4	494	201	516
LEVEL 5	377	951	117
LEVEL 6	145	958	161
LEVEL 7	355	464	207

Response Sheet 2

Taste the samples in the order on the ballot, <u>from left to right</u>. If you have a tray, check to see that the Key numbers match the ones on your samples.

Circle the number of sample in <u>each</u> level is the most different from the other two. If unsure you must guess.

Rinse your mouth with water and wait about 30 seconds before moving to the next Level.

LEVEL 1	281	277	542
LEVEL 2	948	983	986
LEVEL 3	977	299	496
LEVEL 4	356	658	574
LEVEL 5	247	998	816
LEVEL 6	265	300	736
LEVEL 7	693	556	642

Class Data Sheet—Forced choice thresholds

BET	Name or ID	Please mark + under the level(s) that you answered CORRECTLY, 0 for incorrect						
		1	**2**	**3**	**4**	**5**	**6**	**7**
	Proportion Correct							

Sample codes and answer keys for two series. Underlined codes are the odd samples.

Level	Sample Key #1		
1	<u>247</u>	188	633
2	172	962	<u>741</u>
3	657	<u>790</u>	569
4	494	201	<u>516</u>
5	377	<u>951</u>	117
6	<u>145</u>	958	161
7	355	<u>464</u>	207
	Sample Key #2		
1	281	277	<u>542</u>
2	948	<u>983</u>	986
3	<u>977</u>	299	496
4	356	658	<u>574</u>
5	247	<u>998</u>	816
6	<u>265</u>	300	736
7	693	<u>556</u>	642

Signal Detection Theory and the Effect of Criterion on Response

6.1 Instructions for Students

6.1.1 Objectives

To become familiar with the methodology used in signal detection studies.

To gain an understanding of response bias, panelist motivation, and payoff.

To gain an understanding of signal detection as an alternative theoretical framework to classical threshold approaches.

6.1.2 Background

Signal detection theory is an influential framework for conceptualizing discrimination. It is unique from classical threshold approaches such as the method of limits in that it allows the separation of response bias from true discrimination. Signal detection can be used to measure the discriminability of two stimuli or it can be used to measure the resolving power of a panelist. When the discrimination issue is the discernment of a weak stimulus from a blank or baseline item, signal detection can be used as a replacement for threshold theory and measurement. The two versions of the product will be referred to as "target" or signal for the stronger version and "blank" or noise for the weaker one.

Signal detection theory operates under the following assumptions:

1. Sensations arising from presentations of blanks and targets are both normally distributed—sometimes stronger, sometimes weaker—around a mean value. These hypothetical distributions of subjective experiences are called "noise" and "signal" distributions, respectively.

2. Respondents adopt a cutoff value somewhere along the intensity continuum so that one response is given for a stronger sensation (e.g., "I smell it") and a different response is given for a weaker sensation ("I cannot smell it"). Usually a few practice trials are given with feedback to help them choose their cutoff. Note that there are two kinds of presentations (target, blank) and two possible responses, leading to a two by two response matrix.

3. The discriminability of the target from the blank is indicated by the distance between the means of the signal and noise distributions. How this is determined is shown next.

In a signal detection study, then, we can present the target and blank on a blind basis several times (often many, many trials are given) and record the responses.

The discrimination value is determined by the following steps:

1. Count up the proportion of hits, i.e., correct or positive responses when a target trial is given and the proportion of false alarms, responding positively, i.e., incorrectly, when a noise trial is given. These proportions correspond to the hatched areas shown in Fig. 6.1. P(hits)

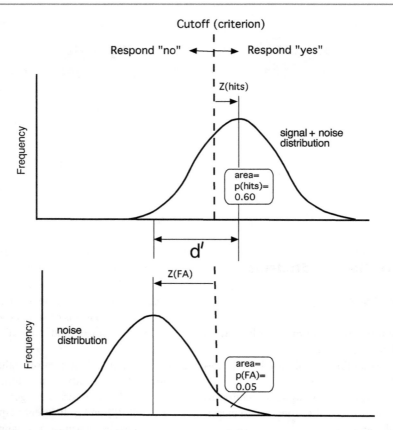

Fig. 6.1 Hypothetical distributions of experiences from signal and noise trials, with a criterion or cutoff drawn in. Experiences stronger than the cutoff generate a "signal" response ("heavier" in this lab) and weaker ones, a noise response ("standard" in this lab). The crosshatched areas correspond to the proportions of hits and false alarms

corresponds to the area under the signal distribution to the right of the cutoff, while P(false alarms) is the area under the noise distribution to the right of the cutoff.

2. Convert the areas to a distance measure in order to find the degree of separation between the means of the two distributions. We convert the proportions of hits and false alarms to z-scores and mathematically "add" them up. Because of the way z-scores are usually tabulated, this is functionally a process of subtraction. The final formula for discriminability without response bias is given by d-prime (d'):

$$d' = Z(\text{hits}) - Z(\text{false alarms})$$

Calculating d-prime is advantageous in that it factors out the response bias or personal criterion that is inherent in a yes/no response method. Individuals might be more conservative or lenient

in how easily they respond "yes" to any stimulus. In the diagram, this tendency would affect the position of the cutoff line. People with a tendency to respond "yes" more often would have a cutoff to the left of those with a tendency to respond "no" more often. However, regardless of the relative position of the cutoff between individuals, the separation between the mean of the noise distribution and the mean of the signal distribution stays the same and d-prime does not change. If a person becomes more lenient, he or she will have more hits but also have more false alarms. If he or she becomes more conservative, the false alarm rate will go down but at the expense of a lower hit rate.

To sum, the cutoff position can slide back and forth, as a person changes his or her criterion or bias. But d-prime remains constant. In the exercise that follows, we will demonstrate the effect of changing the cutoff by manipulating the payoffs for correct responses and the penalties for incorrect responses by changing a payoff matrix.

6.1.3 Materials and Procedures

6.1.3.1 Materials (For Each Pair of Students)

Obtain from your instructor or TA:

Two brown bottles filled with sand or a like material. One weighing a total of 200 g, the other weighing 208 g (The bottles will be identified as standard "S" and heavy "H" only on the bottom.)

$1.75 in change (5 quarters, 7 nickels, and 15 pennies)

A file folder or similar barrier to act as a divider between "experimenter" and "panelist"; payoff matrix keys; two ballots, one for each payoff system; and calculation sheets

6.1.3.2 Procedures

Work in pairs where one student acts as "experimenter" and the other acts as "panelist." When you have finished both payoff systems, you will switch roles and repeat the exercise.

Half of the class (as designated by your instructor or the TAs) will begin with the conservative payoff system, while the other half will begin with the lax payoff system. After you switch roles, begin with whichever payoff system was performed last in your pair.

"Panelist" and "experimenter" should sit across the table or side by side, and the bottles, ballots, and payoff matrices should be set up behind a manila folder so that only the experimenter can see them.

Practice Trials: *P1–P10 on the Ballot*

Begin with the payoff system indicated by your instructor or TA.

Hand the panelist the bottle indicated in trial P1 ("S" or "H") in such a way that she or he cannot see the label on the bottle.

Say to the panelist "this is the standard sample" or "this is the heavier sample" (whichever is appropriate) as you hand each sample to him or her. Allow the panelist a moment to evaluate the weight of the sample.

Take the sample back and proceed to the next trial (through trial P10).

Test Trials: *1–50 on the Ballot*

Allot to the panelist 25 cents (4 nickels and 5 pennies) as his or her opening "stake." Reserve the remaining $1.50 as the "bank."

Hand the panelist the bottle indicated in trial 1 just as you did in the practice trials. Do not tell the panelist the identity of the bottle (i.e., "S" or "H"), but ask her or him to evaluate the weight of the bottle and decide whether this sample is the "standard" or the "heavier" sample. If they are unsure, they must guess.

When the panelist responds, take the bottle back, record the response in the appropriate place on the ballot, and pay the panelist for a correct answer or assess a penalty for an incorrect answer as indicated on the appropriate payoff matrix. Remember to pay or penalize *after every trial*. Do not just tally it up and pay at the end.

Proceed in a similar fashion until you have completed all test trials (through trial 50). Then switch to the next payoff system, and once you have completed the second payoff system, switch roles as "experimenter" and "panelist" and repeat the above procedures for both payoff systems.

When you have finished collecting data, count the number of hits and false alarms:

A hit is when the panelist responds "heavier" when in fact the heavier sample was presented. It does not include all correct responses, just those correct responses when the heavier sample was presented (i.e., positive responses to signal trials).

A false alarm is when the panelist responds "heavier" when in fact the standard sample was presented. It does not include all incorrect responses, just those incorrect responses when the standard sample was presented (i.e., positive responses to noise trials).

Next, determine the proportions of both by dividing the number of hits into the number of "heavier" presentations (25) and by dividing the number of false alarms into the number of "standard" presentations (25). Do not divide by 50!

Convert these proportions to *z*-scores using the table provided, and calculate *d*-prime using the above formula. Spaces for all of these calculations are on the form provided.

Finally, give your calculations to your instructor or TA so they may be checked and recorded. Please return the change to your instructor or TA.

6.1.4 Data Analysis

Convert your proportions to z-scores as discussed above while you are in class. When you have the complete data set or it has been posted, compare the d-prime values, hit rates, and false alarm rates for the two conditions. Do this by performing three paired t-tests. In comparing hit rates and false alarm rates, do the t-tests on the z-scores (not the proportions). The t-tests can be done with a calculator or you may use a statistical program such as Excel. Report the mean values for each condition and the standard errors.

6.1.5 Reporting

Use the standard report format unless instructed otherwise.

6.1.5.1 Results
1. Show the means and standard errors (for all six data columns).
2. What were the outcomes of your three t-tests?
3. Which responses changed (significantly) or did not, and in what direction?

6.1.5.2 Discussion
Answer the following questions for your discussion:
1. How did your *individual* proportions of hits and false alarms change under the two conditions of reward? Was this also true of the class in general (here you can use two of your t-tests)?
2. What does the theory predict for the changes in hits and false alarms as the criterion changes under the different payoff systems? In other words, under what conditions should hits and false alarms go up or down? Did your results agree with the theory?
3. In what ways could this method be applied to an industrial sensory testing situation?

4. What did this experiment tell you about the usefulness of simple yes/no procedures like the one we did here? Do you see any problems or limitations with doing difference tests on food products using this method? What other method(s) could be substituted?
5. Why are blank (noise) trials necessary in signal detection experiments?

Append your calculations to the report.

6.2 For Further Reading

Lawless HT, Heymann H (2010) Sensory evaluation of foods, principles and practices, 2nd ed., Springer Science+Business, New York

Macmillan NA, Creelman CD (1991) Detection theory: a user's guide. Cambridge University Press, Cambridge

O'Mahony M (1992) Understanding discrimination tests: a user-friendly treatment of response bias, rating and ranking R-index tests and their relationship to signal detection. J Sens Stud 7:1–47

6.3 For Instructors and Assistants

6.3.1 Notes and Keys to Successful Execution

1. The exercise with weights was chosen instead of using a food or beverage due to the need to have at least 25 signal and 25 noise trials under each condition. Other options include tactile stimuli such as sandpapers of different grit sizes or a visual difference that is difficult to discern. The stimuli used must be difficult to discriminate, and some pilot work is needed if you do not use the weights recommended here. The items presented can themselves be a bit variable, although the weight jars used here are obviously only present as one (therefore physically identical over trials) member of a pair.
2. If a nonvisual stimulus is used such as the weights, it is important that the two items must not give any clues as to which is which. Otherwise the students may start to base their responses on visual cues rather than the per-

ceived heaviness. It is tempting to "cheat" to get more correct answers. One method for dealing with this is to have the student lift the weighted jars without looking at them. The "experimenter" can pass the jar to the student's side from behind a screen (usually a file folder on its side) and the "subject" student reaches to their side and lifts the jar while facing forward. A blindfold could also be used but is usually not necessary. The exercise is best done in an open classroom with movable chairs, a room with tables or lab with stools and benches, and not a sensory booth facility.

3. Because of the need to change payoff matrices, this lab can take over an hour or two to complete. Different students will work at different paces. It is recommended to allow students to leave after their data are tabulated, rather than having to wait for the slowest pair to finish.

4. Several rolls of coins are needed for the payoffs. If a student "breaks the bank," they can be issued IOUs. Conversely, if they perform so poorly as to lose all their stakes, then they can be given a bank loan. Performance can be motivated by giving a token gift to the highest performing student(s), i.e., the highest d-prime(s) obtained. An alternative to coins is fake paper money from a board game such as Monopoly.

5. It is important to give the student feedback *after every trial*. One common mistake is for the experimenter/student to just record the performance and calculate the amount won or owed at the end, to make things easier. This is a fatal mistake as it does not induce the student to change his or her criterion as the test progresses. TAs should observe each student pair to make sure money is changing hands on every single trial, after the practice runs.

6. Some students may not be able to discriminate the weights. This possibility should be mentioned to the class to prevent embarrassment. They should be encouraged to try to go through the exercise. Sometimes they will feel they cannot do the task, but their performance often indicates otherwise.

7. It is important to distinguish the payoff matrices from the actual criterion levels. A payoff scheme matrix is neither lax nor conservative; it is structured to attempt to change the person's criterion or cutoff from lax to conservative or vice versa. Hits and false alarm rates should go up or down together. This may not be seen in every student's data, but this lab does work well on a class average basis. The t-tests should show a shift in the proportions and z-scores, but not the d' values. To deal with any practice effect, half the class should start with one payoff matrix and the other half with the second one.

8. It is recommended to have students lift with the hand by grasping the top of the jar and making two or three up and down motions.

9. The tedium of this procedure should also raise the question whether the extended yes/no task is appropriate for an industrial sensory testing method. You can draw the parallel to the A-not A test and emphasize that d' values can now be obtained from other discrimination tests as discussed in Chap. 5 in Lawless and Heymann (2010).

6.3.2 Equipment

Weighing balance(s).

6.3.3 Materials and Supplies

Four ounce amber jars (or similar) with lids.
Material to fill jars to the appropriate weight.
$1.75 per student pair is required in mixed change.
A table for converting proportions to z-scores.

6.3.4 Procedures (See Notes and Supply List)

Jars should be filled to 200 g (standard) and 208 g (signal, heavier).
Jars may be filled with sand, sugar, or some other easily handled granular material (such BB's). Sugar, if kept clean, may be reused for subsequent lab exercises. Jars must be physically and visually indistinguishable and coded on the bottom out of sight. If the level of fill is perceivable, jars may be covered with

aluminum foil or a piece of stiff paper may be curled and inserted into the jar before filling to conceal the amount of fill.

Forms: Response sheets, payoff matrix keys, tabulation sheets for individual students, and tab sheets for class data are needed.

It is useful to have an overhead or slide of the response key, to remind them that a hit occurs when the response is "heavier" to the 208-g weight (H for heavy) and a false alarm is when the response is "heavier" to the 200-g weight (S for standard).

6.4   Appendix: Ballots and Forms

6.4.1   Individual Ballots

Trial	Pres.	Resp.	Trial	Pres.	Resp.	Trial	Pres.	Resp.
P1	H		11	S		31	S	
P2	S		12	S		32	H	
P3	H		13	H		33	H	
P4	S		14	S		34	S	
P5	H		15	S		35	S	
P6	S		16	H		36	S	
P7	H		17	S		37	H	
P8	S		18	S		38	H	
P9	H		19	S		39	H	
P10	S		20	H		40	S	
1	H		21	S		41	S	
2	S		22	H		42	S	
3	H		23	H		43	S	
4	H		24	H		44	H	
5	S		25	S		45	S	
6	H		26	H		46	H	
7	H		27	H		47	S	
8	S		28	S		48	H	
9	H		29	H		49	S	
10	H		30	S		50	H	

SYSTEM: _____ **LAX**   _____ **CONSERVATIVE**

Trial	Pres.	Resp.	Trial	Pres.	Resp.	Trial	Pres.	Resp.
P1	H		11	S		31	S	
P2	S		12	H		32	H	
P3	H		13	H		33	S	
P4	S		14	S		34	S	
P5	H		15	S		35	H	
P6	S		16	H		36	S	
P7	H		17	H		37	S	
P8	S		18	H		38	S	
P9	H		19	S		39	H	
P10	S		20	H		40	S	
1	S		21	S		41	S	
2	S		22	H		42	S	
3	H		23	S		43	H	
4	H		24	H		44	H	
5	S		25	S		45	S	
6	H		26	H		46	H	
7	S		27	H		47	S	
8	H		28	S		48	H	
9	H		29	H		49	H	
10	H		30	S		50	S	

SYSTEM: _____ **LAX**   _____ **CONSERVATIVE**

Trial	Pres.	Resp.	Trial	Pres.	Resp.	Trial	Pres.	Resp.
P1	H		11	H		31	H	
P2	S		12	S		32	S	
P3	H		13	H		33	S	
P4	S		14	S		34	S	
P5	H		15	S		35	H	
P6	S		16	H		36	S	
P7	H		17	H		37	S	
P8	S		18	S		38	H	
P9	H		19	S		39	S	
P10	S		20	H		40	S	
1	S		21	S		41	H	
2	H		22	H		42	S	
3	S		23	S		43	H	
4	H		24	H		44	H	
5	S		25	H		45	S	
6	H		26	H		46	H	
7	H		27	S		47	S	
8	S		28	H		48	S	
9	H		29	S		49	H	
10	H		30	S		50	H	

SYSTEM: _____ **LAX**   _____ **CONSERVATIVE**

Trial	Pres.	Resp.	Trial	Pres.	Resp.	Trial	Pres.	Resp.
P1	H		11	H		31	H	
P2	S		12	S		32	S	
P3	H		13	H		33	H	
P4	S		14	S		34	S	
P5	H		15	S		35	H	
P6	S		16	H		36	S	
P7	H		17	H		37	S	
P8	S		18	H		38	H	
P9	H		19	S		39	S	
P10	S		20	H		40	S	
1	H		21	S		41	H	
2	H		22	S		42	S	
3	S		23	S		43	H	
4	S		24	H		44	H	
5	H		25	H		45	S	
6	S		26	H		46	S	
7	H		27	S		47	S	
8	S		28	H		48	H	
9	H		29	S		49	H	
10	H		30	S		50	S	

SYSTEM(check):   _____ **LAX**   _____ **CONSERVATIVE**

6.4.2   Class Data Sheet

Class Data Sheet

NAME	CONSERVATIVE			LAX		
	z_H	Z_{FA}	d'	z_H	Z_{FA}	d'
MEAN						

6.4.3   P and z-Score Conversion Table

p- and Z-scores for calculation of d'

p	Z-score	p	Z-score	p	Z-score	p	Z-score
0.01	-2.33	0.26	-0.64	0.51	+0.03	0.76	+0.71
0.02	-2.05	0.27	-0.61	0.52	+0.05	0.77	+0.74
0.03	-1.88	0.28	-0.58	0.53	+0.08	0.78	+0.77
0.04	-1.75	0.29	-0.55	0.54	+0.10	0.79	+0.81
0.05	-1.64	0.30	-0.52	0.55	+0.13	0.80	+0.84
0.06	-1.55	0.31	-0.50	0.56	+0.15	0.81	+0.88
0.07	-1.48	0.32	-0.47	0.57	+0.18	0.82	+0.92
0.08	-1.41	0.33	-0.44	0.58	+0.20	0.83	+0.95
0.09	-1.34	0.34	-0.41	0.59	+0.23	0.84	+0.99
0.10	-1.28	0.35	-0.39	0.60	+0.25	0.85	+1.04
0.11	-1.23	0.36	-0.36	0.61	+0.28	0.86	+1.08
0.12	-1.18	0.37	-0.33	0.62	+0.31	0.87	+1.13
0.13	-1.13	0.38	-0.31	0.63	+0.33	0.88	+1.18
0.14	-1.08	0.39	-0.28	0.64	+0.36	0.89	+1.23
0.15	-1.04	0.40	-0.25	0.65	+0.39	0.90	+1.28
0.16	-0.99	0.41	-0.23	0.66	+0.41	0.91	+1.34
0.17	-0.95	0.42	-0.20	0.67	+0.44	0.92	+1.41
0.18	-0.92	0.43	-0.18	0.68	+0.47	0.93	+1.48
0.19	-0.88	0.44	-0.15	0.69	+0.50	0.94	+1.55
0.20	-0.84	0.45	-0.13	0.70	+0.52	0.95	+1.64
0.21	-0.81	0.46	-0.10	0.71	+0.55	0.96	+1.75
0.22	-0.77	0.47	-0.08	0.72	+0.58	0.97	+1.88
0.23	-0.74	0.48	-0.05	0.73	+0.61	0.98	+2.05
0.24	-0.71	0.49	-0.03	0.74	+0.64	0.99	+2.33
0.25	-0.67	0.50	0.00	0.75	+0.67	0.995	+2.58

6.4.4   Response Keys and Individual Response Calculation Sheets

Payoff matrix keys and individual calculation sheets

1. Payoff system designed to induce a conservative criterion:

	Response	
	"yes/heavier"	"no/standard"
Heavier/signal presented	+1 (HIT)	-1
Noise (standard) presented	-5 (False alarm)	+5

Student Total:

Hits_____ (divide by 25 to get proportion)
Proportion:_____ Z-score_____

False alarms:_____ (divide by 25 to get proportion)
Proportion:_____ Z-score_____

2. Payoff system designed to induce a conservative criterion:

	Response	
	"yes/heavier"	"no/standard"
Heavier/signal presented	+5 (HIT)	-5
Noise (standard) presented	-1 (False alarm)	+1

Student Total:

Hits_____ (divide by 25 to get proportion)
Proportion:_____ Z-score_____

False alarms:_____ (divide by 25 to get proportion)
Proportion:_____ Z-score_____

Sweetness of Fructose and Sucrose Determined by Different Scaling Methods

7

7.1 Instructions for Students

7.1.1 Objectives

To gain familiarity with two common scaling methods: magnitude estimation and line scaling (in some years, category scaling).

To examine the psychophysical functions that describe stimulus-response relationships for these two methods.

To understand the importance of units of measurement when drawing conclusions about flavor potency.

To gain experience in scientific graphing and construction of figures.

7.1.2 Background

Scaling Methods. Various methods have been used in sensory science that apply numbers to reflect changes in the perceived intensity of a sensation (Lawless and Malone, 1986). A common method used in food science is the simple category scale. In this method, a category response is chosen such as an integer number or a check box and the data are assumed to be used in a linear fashion. By "linear" we mean that equal number/scale differences represent equal differences in perceived intensities. Another common method is to have intensities rated using a line scale to represent equal steps in perceived strength of an attribute. These are sometimes called "visual analogue scales" or VAS

for short. An alternative approach to the line or line scale is to rate perceived intensities relative to a standard using a ratio or proportional method. Equal ratios between the values used to describe intensities (e.g., 1–2 or 5–10) reflect equal proportions in the relative perceived intensities. This type of scaling is called magnitude estimation. Category scaling and magnitude estimation have been criticized for judges' nonlinear use of numbers. Category and line scale data often correspond to a log relation (i.e., yields a straight line when plotted as a function of log stimulus concentration, as in Fechner's law) (Baird and Noma, 1978), but magnitude estimation data are usually fit by a power function (shown by a straight line when log responses are plotted as a function of log stimulus concentration, as in Steven's law) (Stevens and Galanter, 1957).

Relative Potency of Sweeteners. Many different claims have been made in the food science literature concerning the relative sweetness potencies of various sugars and high-intensity sweeteners (Cardello et al., 1979). Sometimes these claims are based on different units of measurement. A monosaccharide such as fructose or glucose might be sweeter than a disaccharide such as sucrose on a weight-to-weight basis. If a food processor or manufacturer who buys sweetener by weight were interested in various levels of sweetness, this would represent an appropriate means of comparison. But, if a biochemist were interested in the binding process between molecules and taste receptors that leads to the sensation of sweetness, a comparison based on molarity (i.e., molecules per unit volume) would be more appropriate. Equal molarities have equal numbers of molecules in a given volume of sample. Changing the units of comparison makes a big difference! Should you equate grams or molecules for a fair comparison? The answer may depend upon whether you think like a chemist or a food manufacturer.

In the following exercises, you will compare the relative sweetness impact of sucrose and fructose on both a weight per unit volume and a molar basis. You will also be asked to make eight graphs. The ability to present data in clear manner is key to scientific communication. Finally, you will compare the functions from the two scaling methods to see which mathematical function seems to fit better (log or power function).

See Chap. 7 in Lawless and Heymann (2010) for further information.

7.1.3 Materials and Procedures

7.1.3.1 Materials
Samples will consist of a powdered fruit beverage sweetened with sucrose at the following concentrations: 2.5% w/v(0.073 M), 5.0 % w/v(0.146 M), 10.0 % w/v(0.292 M), and 20.0% w/v(0.585 M). The same product will be made with fructose at the following concentrations: 2.5% w/v(0.139 M), 5.0% w/v(0.278 M), 10.0% w/v(0.556 M), and 20.0% w/v(1.111 M).

7.1.3.2 Procedures
(Note: In some years, a category scale may be used in place of a line scale.)

Rate the sweetness of each of the above eight solutions using both a line scale and a magnitude estimation scale with the 5.0 % w/v sucrose solution used as a standard (assigned a value of "10"). The standard will be tasted *before* the first, third, fifth, and seventh samples in the magnitude estimation exercise.

All ratings should be made in one scale type before moving to the other. Half of the class should begin with the line scaling method and the other half should begin with magnitude estimation. Students may form pairs and switch places

as experimenter and panelist when they are done. The TA will fix sample presentation orders so that the orders of presentation within a sweetener are randomized.

Once ratings have been completed, enter all data onto master sheets for both scaling methods. You will calculate means, standard deviations, and standard errors from these data. Spreadsheets containing these data will be sent to you by e-mail or posted on the class website, along with a key of sample codes and sugar concentrations. You will use the means and standard errors to make various plots of sweetness vs. concentration.

7.1.4 Data Analysis

Compute means and standard errors for the four concentrations of each sugar and each scaling method. You should have data for four curves or lines to plot.

7.1.5 Reporting

Format: Your report will consist of eight graphs and a short discussion that answers all the discussion questions in complete English sentences. See the notes below on graphing hints.

Construct the following graphs from the above data with the variables for X and Y axes as indicated. Note that the first four have normal arithmetic scales for the axes and the second set of four have log-transformed axes on the X variable or both X and Y. You should include error bars if possible on the first four graphs and always plot sucrose and fructose data on the same graph.

(a) X: weight per vol. conc., Y: line scale ratings
(b) X: weight per vol. conc., Y: magnitude estimation ratings
(c) X: molarity, Y: line scale ratings
(d) X: molarity, Y: magnitude estimation ratings
(e) X: log molarity, Y: line scale ratings
(f) X: log molarity, Y: magnitude estimation ratings
(g) X: log molarity, Y: log line scale ratings
(h) X: log molarity, Y: log magnitude estimation ratings

Place both sweeteners on each graph so the relative intensities can be compared. Semilog graph paper may be used for the semilog plots (e and f), and log-log graph paper may be used for the log-log plots (g and h), or you may use the "log axis" options in a graphing program or Excel. Do not do both! (You would be log transforming twice!)

Include standard error bars on the arithmetic plots (a–d). The standard error equals the standard deviation divided by the square root of N (number of panelists or observations).

Answer and discuss the following questions:

1. Which sugar is sweeter, sucrose or fructose? Explain your reasoning.
2. How does the change in units of measurement change the relationship between the two sugars?

3a. Which unit of measurement is more useful from a business point of view? Why?

3b. Which makes more sense from a biochemical point of view? Why?

4. How could you express the relative potencies of the two sugars? (There is more than one way to do this—consider a concentration ratio.)

5. What type of psychophysical function best fits each scaling method? What is predicted by theory? Did your graphs agree with the theory? Is this visible in the plots (state which plots and why)?

7.1.6 Graphing and Hints

Your grade is partly determined by the graphs. You may use a graphing program or you may draw them by hand, but please do so *neatly* (use a ruler to make straight lines). Examples may be shown in class.

If using a plotting program, it is very important to not simply accept the default options in any program like Excel. The default options may produce graphs with odd choices and fonts that are difficult to read. To change the defaults, use the help function if you do not know exactly what to do. Make additional adjustments beyond the program defaults to make your graphs communicate well. Follow the guidelines listed below as appropriate. Making good graphs will take some

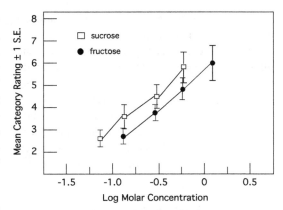

Fig. 7.1 Semi-log plot of concentration vs. mean perceived sweetness for fructose and sucrose

time. Do not leave your graphs until the last minute. An example of an acceptable graph is shown in Fig. 7.1.

Always label the axes and at least some of the tick marks (numbers). State your units. Indicate if axis is plotting logs or if a log scale is used.

The number range on a Y-axis should start at a value close to the response minimum and end about 30 % more than the maximum in the data. Do not start with zero if the response scale starts at 1.

Beware of extra and unnecessary decimal points and zeroes. Excel may give you lots of zeroes. Do not accept this—change it! Limit zeros to one or two decimal places for the axis tick labels.

Generally, the number of decimal points should not exceed the precision of your data. If your standard errors are ±0.5 units, you should have only one place beyond the decimal.

The axes and their labels should be placed on the bottom and left-hand side, not in the middle of a graph. Excel may put them wherever zero occurs. Do not accept this—change it!

Use symbols that are big enough to see the difference. Use large legible fonts (12 or 14 point) for lettering on axis labels and numbers.

Avoid using color as a code for different curves or bars. Many scientific journals still publish graphs in black and white.

Do not shade or fill the graph to provide a gray background. It obscures the data.

Include a legend or key to label different curves.

Include a caption or title that completely describes what is plotted.

Do not squeeze axis legends into small margins. Leave space around the graph box for the text.

Avoid graphs that are too short and wide or too narrow and high. The eye sees relationships in the curves better when the axes are about equal. A rule of thumb is to use the "golden rectangle" of ancient Greece (about a 2:3 ratio of height to width).

7.2   For Further Reading

Baird JC, Noma E (1978) Fundamentals of scaling and psychophysics. Wiley, New York

Cardello AV, Hunt D, Mann B (1979) Relative sweetness of fructose and sucrose in model systems, lemon beverages and white cake. J Food Sci 44:748–751

Lawless HT, Heymann H (2010) Sensory evaluation of foods, principles and practices, 2nd ed., Springer Science+Business, New York

Lawless HT, Malone GJ (1986) A comparison of scaling methods: Sensitivity, replicates and relative measurement. J Sens Stud 1:155–174

Pangborn RM (1963) Relative taste intensities of selected sugars and organic acids. J Food Sci 28:726–733

Stevens SS, Galanter EH (1957) Ratio scales and category scales for a dozen perceptual continua. J Exp Psychol 54:377–411

7.3   For Instructors and Assistants

7.3.1   Notes and Keys to Successful Execution

1. In making sample solutions, it is important to instruct the preparers that they cannot simply add a fixed weight of sugar to a fixed volume of water. The sugar will cause the volume to expand, creating an inaccurate concentration. Instead, the desired weight of sugar can be added to about 50–75 % of the final desired concentration and mixed thoroughly using a stir plate or similar device. Once the sugar is dissolved (this may take some time), the solution can be "topped off" to get to the final desired volume, with continued mixing. Large volumetric flasks are recommended.

2. If prepared ahead of time, refrigerate for safety, then bring samples to room temperature for serving.

3. The higher concentrations may take some patience in getting that much sugar into solution. Stir, but be careful of heating so as not to create invert sugar from sucrose.

4. Do not underestimate the time needed for pouring samples. Students can be recruited to assist the preparers for this step.

5. If you do not tell them the molarities, you can check that they can covert wt/vol to moles per liter. This will reinforce some basic chemistry. I am amazed at the number of students who are stumped by this simple conversion. MW of sucrose is 342 and MW of fructose is 180. Obviously, you lose a water molecule (=18) when making the disaccharide, so it is not quite twice as heavy.

6. You can use line scales instead of category scales in alternate years to help discourage copying from a former student or from files that may persist on campus.

7. Some students may suggest that sucrose hydrolyzes, and so it is sweeter because there are twice as many molecules. To the best of our knowledge, the hydrolysis is very slow and thus negligible. That is, you do not get "invert sugar" overnight, unless you add acid and heat the solution. If you use a commercial powdered drink mix and serve it within 24 h, the hydrolysis should not be a factor.

8. As with many of our other lab exercises, you can decide how much prep work you want the students to participate in and how much data analysis they do (or that you do for them).

9. Most will want to use Excel as a graphing program, but often the default values produce poor-looking graphs. We do not teach Excel but refer them to the help function, a useful habit. Placing error bars on the graphs is a little obscure in Excel. For students who decide to graph by hand (if allowed), one option is to use semilog and log-log graph paper, available in your college bookstore or engineering department or downloadable from the web.

10. What constitutes a straight line when looking
for the curve fits? Some students may think
of fitting a line to their graphs, which is easy
to do in Excel, and looking at the R-squared
value. You can require this as part of their
analysis, but the fits are generally pretty high
anyway. Another option is just to have them
look "by eye." Note: there are often deviations from a linear function at higher concentrations (due to saturation of receptors and
thus response) and also at the lower end (you
bump into threshold at some point). The latter should not be a problem since the starting
level is well above the sweetness threshold.

11. This is a complex lab exercise with four
parts: scaling methods, sweetness issues,
graphing, and psychophysical functions. It
may not be suitable for all students depending upon the level at which sensory is offered.
It is possible to simplify by omitting the log
axis graphs and also the questions dealing
with which psychophysical functions and
curve fitting.

12. You can use the experimenter-subject format
where students work in pairs, or they can
serve themselves from trays, but make sure
they follow some random orders.

13. Further background. Fructose is widely
touted by suppliers as sweeter than sucrose,
so you can use less in any given food appli-
cation and save on costs. This agrees with the
"wisdom" of most food chemistry texts.
However, the pattern is much more compli-
cated. Fructose potency depends upon the
other ingredients present, the pH, and the
temperature. Fructose in water exists in an
equilibrium between its furanose and pyra-
nose forms (5 and 6 atom rings), and the
sweeter version is favored at lower temperatures. See the paper by Pangborn (1963) to
see how the relative sweetness changes once
the fructose is presented in pear nectar. The
advantage disappears.

On a weight/volume basis, the students'
data usually agree with the literature, that
fructose is slightly "sweeter," by which we
really mean that it is *more potent* (higher
response for a given concentration). On a

molar basis, however, sucrose is far sweeter
than fructose. The potency is shown by the
relative height of the two curves at any given
concentration. Many students will want to
choose a given concentration and compare
the response. This is the opposite of what is
normally done in flavor science. The common approach is to take an iso-sweet (equal
response level) and compute the concentration ratio. This is how we come up with the
somewhat nonsensical statements like "sac-
charin is 200 times sweeter than sucrose."
Actually it is not. You just use 1/200th as
much to get the same response. One common
iso-sweet level is the threshold. Another,
popular in foods and beverages, is to look at
the 10 % sucrose equivalent. They actually
have this data point.

Many taste researchers would consider a
weight-based comparison of two substances
as flawed or at least not relevant. That is
because they are thinking like biochemists,
who would prefer to compare molecule to
molecule rather than gram to gram. So a
molar comparison makes sense to someone
working in the area of chemical interactions
of taste molecules with receptors.

Why is sucrose sweeter on a molar basis?
You can discuss the AH-B theory (hydrogen
bonding) of sweetness and suggest that it
may have more AH-B binding sites than
fructose (after all it is a disaccharide with
more hydroxyl groups per molecule). You
can also reinforce the notion that hydrogen
bonds (energetically fast and loose) are common in physiological systems. Students
should remember something about H bonds
from hearing about DNA base pairs.

14. Fitting psychophysical functions (dose–
response curves). If students examine the
log-log and semilog graphs, they can look
for a straight line to see if the dose–response
function fits "Fechner's Law," i.e.,
$R = k \log(C)$ where R is response and C is
concentration, or "Stevens Law," $R = k(C)^n$
and thus $\log(R) = n \log(C) + \log(k)$ where
n is the characteristic exponent of the magnitude estimation function. This obviously

becomes the slope in a log-log plot. According to the literature, the log function often fits category scale data and the power function fits magnitude estimation data.

15. See "common errors in graphing" below for assistance in grading or suggestions to write on graphs.

7.3.2 Equipment

Weighing balance(s), stir plates, stir bars, and volumetric flasks

7.3.3 Supplies

Powdered drink mix. The lab can also be done in aqueous solutions, but it is not much fun to taste sugar in plain water.

Spring water or odorless water of higher purity.

Sample cups, rinse cups, spit cups, napkins, materials for spill control, and trash disposal. Remember you will need approximately 20 sample cups per student.

7.3.4 Procedures

See notes above on preparing solutions. It is important to start with less water than the desired final amount because the sugars will expand the volume of the liquid they are dissolved in. Once dissolved, the samples can be topped off while mixing to arrive at the correct final volume.

For samples to taste, generally 20 ml is a sufficient sample volume. It is useful to have a small tray for each student setup with the coded samples in the correct randomized orders if possible. Alternatively, they can be provided with a key or be told to follow the order printed on the ballots.

Remember that the magnitude estimation sets will require three extra samples of 5 % wt/vol sucrose, which can be labeled simply "REF."

Sample keys are shown in tables in the appendix before the sample ballots.

7.3.5 Grading Suggestions: Common Errors in Graphing

Failure to label graphs and/or axes. What is plotted (means?)?

Symbols too small.

Failure to provide a caption or descriptive title.

Failure to define error bars (SD or SE?).

Using the log option in graphing but labeling it as log concentration. Students must distinguish between log concentration and a plotting on a log scale (pushing the radio button for a log scale axis).

Logging the wrong axis in semilog.

Too many decimal places in the plots or axis or tick-mark labels.

Failure to provide a descriptive legend or key (sucrose, not "series 1").

7.4   Appendix: Sample Ballots, Keys, and Data Sheets

MAGNITUDE ESTIMATION

Taste the REFERENCE sample (marked "REF"). Assign it a SWEETNESS rating of 10.

Next, taste each of the test samples in the order as listed below. Assign each test sample a SWEETNESS rating relative to the REFERENCE sample.

(For example, if you perceive a test sample to be twice as sweet as the Reference sample, assign it a rating of 20; if the test sample is half as sweet as the Reference sample, assign it a rating of 5).

You may use any positive integer or fraction as your rating.

Please re-taste the Reference sample as you are instructed to do so.

                                                         Sweetness Rating
Please taste the REFERENCE sample

Sample 197   _____

Sample 402   _____

Please re-taste the REFERENCE sample

Sample 493   _____

Sample 716   _____

Please re-taste the REFERENCE sample

Sample 285   _____

Sample 349   _____

Please re-taste the REFERENCE sample

Sample 586   _____

Sample 362   _____

MAGNITUDE ESTIMATION

Taste the REFERENCE sample (marked "REF"). Assign it a SWEETNESS rating of 10.

Next, taste each of the test samples in the order as listed below. Assign each test sample a SWEETNESS rating relative to the REFERENCE sample.

(For example, if you perceive a test sample to be twice as sweet as the Reference sample, assign it a rating of 20; if the test sample is half as sweet as the Reference sample, assign it a rating of 5).

You may use any positive integer or fraction as your rating.

Please re-taste the Reference sample as you are instructed to do so.

Sweetness Rating

Please taste the REFERENCE sample

Sample 362   _____

Sample 716   _____

Please re-taste the REFERENCE sample

Sample 285   _____

Sample 349   _____

Please re-taste the REFERENCE sample

Sample 493   _____

Sample 402   _____

Please re-taste the REFERENCE sample

Sample 586   _____

Sample 197   _____

LINE SCALING

Taste the following samples in the order that they are presented on this page and mark on the line to indicate the best level of SWEETNESS for each sample.

Sample 515

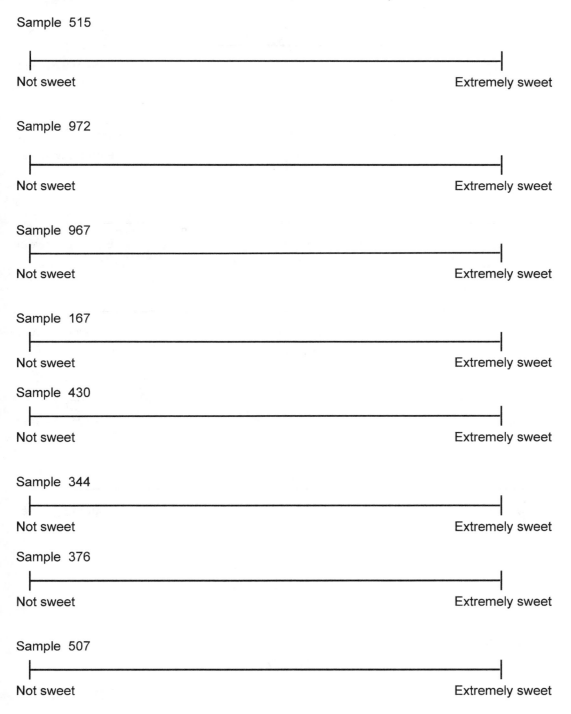

Not sweet   Extremely sweet

Sample 972

Not sweet   Extremely sweet

Sample 967

Not sweet   Extremely sweet

Sample 167

Not sweet   Extremely sweet

Sample 430

Not sweet   Extremely sweet

Sample 344

Not sweet   Extremely sweet

Sample 376

Not sweet   Extremely sweet

Sample 507

Not sweet   Extremely sweet

LINE SCALING

Taste the following samples in the order that they are presented on this page and mark on the line to indicate the best level of SWEETNESS for each sample.

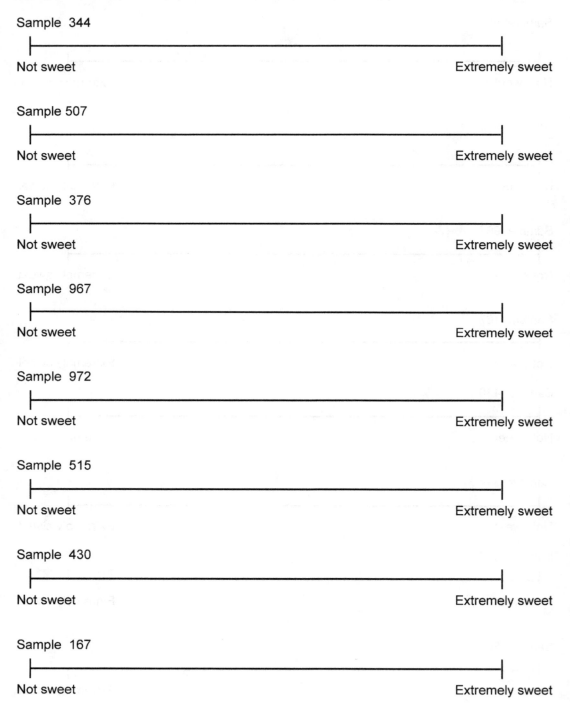

Sample 344

Not sweet   Extremely sweet

Sample 507

Not sweet   Extremely sweet

Sample 376

Not sweet   Extremely sweet

Sample 967

Not sweet   Extremely sweet

Sample 972

Not sweet   Extremely sweet

Sample 515

Not sweet   Extremely sweet

Sample 430

Not sweet   Extremely sweet

Sample 167

Not sweet   Extremely sweet

Magnitude Estimation—STUDENT DATA SHEET

Please record your data in the table below.

	Magnitude Estimation							
	Sucrose				Fructose			
w/v conc	2.5%	5.0%	10.0%	20.0%	2.5%	5.0%	10.0%	20.0%
Molarity	0.073	0.146	0.292	0.585	0.139	0.278	0.556	1.111
Log Molarity	-1.137	-0.836	-0.535	-0.233	-0.857	-0.556	-0.255	0.046
Student ID	716	402	349	586	493	285	197	362
Mean								
Log Mean								
Std Deviation								

Line Scaling—STUDENT DATA SHEET

Please record your data in the table below.

	Line Scaling (cm)							
	Sucrose				Fructose			
w/v conc	2.5%	5.0%	10.0%	20.0%	2.5%	5.0%	10.0%	20.0%
Molarity	0.073	0.146	0.292	0.585	0.139	0.278	0.556	1.111
Log Molarity	-1.137	-0.836	-0.535	-0.233	-0.857	-0.556	-0.255	0.046
Student ID	**967**	**430**	**167**	**376**	**515**	**972**	**344**	**507**
Mean								
Log Mean								
Std Deviation								

Sample key for Magnitude Estimation

w/v Concentration	Sucrose		Fructose	
	Molarity	Code	Molarity	Code
2.5%	0.073	716	0.139	493
5.0%	0.146	402	0.278	285
10.0%	0.292	349	0.556	197
20.0%	0.585	586	1.111	362

Sample key for Line Scaling

w/v Concentration	Sucrose		Fructose	
	Molarity	Code	Molarity	Code
2.5%	0.073	967	0.139	515
5.0%	0.146	430	0.278	972
10.0%	0.292	167	0.556	344
20.0%	0.585	376	1.111	507

Time-Intensity Scaling

8

8.1 Instructions for Students

8.1.1 Objectives

To become familiar with time-intensity methods.
To demonstrate how differences between products are specified by time-intensity curve parameters.
To demonstrate the use of a paper ballot and computerized system for collection of time-intensity data (optional).

8.1.2 Background

Perceptions of aroma, taste, flavor, and texture in foods are dynamic phenomena. The perceived intensity of the sensory attributes change from moment to moment, due to processes of chewing, breathing, salivation, tongue movements, and swallowing. In texture evaluation, different phases of food breakdown are often separated into first bite, mastication, and residual phases. The time profile of a food or beverage can be an important aspect of its sensory appeal.

Common methods of sensory scaling ask the panelists to rate the perceived intensity of the sensation by giving a single (uni-point) measurement. This task requires that the panelists must "time-average" their changing sensations or to estimate only the peak intensity. This can miss some important information, such as a lingering taste that consumers may dislike. It is possible,

for example, for two products to have the same descriptive specifications but differ in the order in which different flavors occur or when they reach their peak intensities.

Time-intensity (TI) sensory evaluation is an attempt to provide the panelists with the opportunity to report their perceived sensations over time. Because the judges are repeatedly (in some cases continuously) monitoring their perceived sensations, the sensory scientist is able to quantify the perceptual changes that occur in the specified attribute over time. When multiple attributes are tracked, the profile of a complex food flavor or texture may show differences between products that change across time after ingestion. For most sensations, the perceived intensity increases and eventually decreases, but for some, like perceived toughness of meat, the sensation only decreases as a function of time. For perceived melting, the sensation of thickness or firmness may only decrease until a completely melted state is reached. The additional information derived from time-intensity scaling is especially useful when studying sweeteners or products like chewing gums, hand lotions, etc., that have a distinctive time profile.

When performing a TI study, the sensory specialist can obtain the following information for each sample and for each panelist: the maximum intensity perceived (Imax), the time to maximum intensity (Tmax), the rate and shape of the increase in intensity to the maximum point, the rate and shape of the decrease in intensity to half maximal intensity and to the extinction point, the

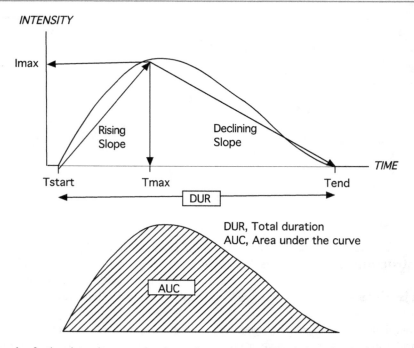

Fig. 8.1 Example of a time-intensity curve showing various parameters that characterize the data

total duration of the sensation, (DUR) and the area under the curve (AUC) as shown in Fig. 8.1.

If your sensory facility has a computer-aided system for data collection, in this exercise you can compare two common methods for time-intensity scaling: a verbally cued "stopwatch" method and a computerized continuous tracking procedure.

Additional background readings may be found in the reference list.

8.1.3   Materials and Procedures

8.1.3.1 Materials
Cinnamon-flavored chewing gum, Tabasco sauce or similar product (room temperature), Tabasco sauce (chilled), unsalted crackers, rinse water.

8.1.3.2 Procedures
Each student will complete time-intensity exercises for each of three products using both a paper ballot and a computerized data system if one is available. If the class is doing both paper ballots and computer data capture, students

should divide into two equal-sized groups at the beginning of the lab session. Half of the students should proceed to the sensory lab and perform their ratings using the computer system, while the other half may remain in the classroom and perform their ratings using paper ballots. When students have finished their ratings, the groups switch places.

1. Practice Exercise: Place the stick of chewing gum in your mouth and begin to chew it. Continue chewing the gum for the entire minute of the rating period.

   The instructor or teaching assistant will tell you when to start. If you are using *paper ballots*, make your first rating of cinnamon intensity the moment you begin to chew the gum and make subsequent ratings when the instructor or teaching assistant tells you to rate. If you are using a computer system, follow the specific instructions on the computer screen.

2. Hot Sauce Exercises: Place three drops of either chilled or room temperature hot sauce on a plastic spoon and taste the entire sample. Rate the heat intensity of the sauce using the appropriate method as described above. The

time intervals may be different than those used for the practice gum.

Wait at least 3 min after the first rating period or until your mouth feels comfortable again. You may cleanse your mouth with water and/or consume crackers if you wish during this interval. At the conclusion of the rest period, rinse your mouth to be sure that your mouth is free from cracker particles and repeat the above procedures, this time using the sauce kept at the other temperature.

8.1.4   Data Analysis

For the paper ballot results, find the individual Tmax (time to maximum), Imax (intensity at maximum), DUR (total duration), and AUC (area under the curve) measures for each student. If there is a plateau or multiple peaks, record the Tmax as the first time at which the peak intensity is reached. If the intensity is not zero at the final period, record the DUR as the final time. For estimating the AUC, you may use triangular approximations for data that show a clear peak (usually two triangles are necessary, one for the rising and one for the falling phase) or a trapezoid if there appears to be a plateau (two triangles and a rectangle under the plateau). An alternative method for measuring AUC is to cut out the curve boundaries using a scissor and weigh the paper. AUC may then be expressed in grams. Perform a paired t-test on each of the parameters you extracted in order to compare the two temperatures of the hot sauce.

Plot a time graph of the class-averaged data for each of the two temperatures.

If your computer systems also provide individual estimates of these parameters, you may also perform the same t-tests for the computer-captured data. If it provides only group estimates, you may compare them qualitatively (i.e., without a statistical test). Show a time graph of the computer-captured class averages for the two temperatures (this may be provided to you, depending upon your system).

8.1.5   Report

Results will be sent to you by e-mail or posted on the course website.

Write the report with the following sections (1) Objectives (what were the goals of the study), (2) Methods, (3) Results, and (4) Discussion. You have a *four-page limit*, not counting graphs. Please type double spaced.

Results should answer the following questions: Did the two temperatures give different results? In what way?
Did the two methods (paper vs. computer) differ? In what way?
Refer to the graphs (e.g., "Fig. 8.1 shows…") in supporting your conclusions. Use any statistics from your t-tests or any that may be provided from the computer analysis to help support your conclusions.

These questions can be answered in terms of the parameters of the TI curves such as Imax, Tmax, DUR, and AUC.

Your discussion should answer the following questions:
Which method, paper ballots or computer tracking, seemed better to you? Why?
Were there any problems encountered such as data censoring or truncation?
What other product systems or food research questions would the TI method be useful for?

8.2   For Further Reading

Gwartney E, Heymann H (1995) The temporal perception of menthol. J Sens Stud 10:393–400

Lallemand M, Giboreau A, Rytz A, Colas B (1999) Extracting parameters from time-intensity curves using a trapezoid model: the example of some sensory attributes of ice cream. J Sens Stud 14:387–399

Lee WE, Pangborn RM (1986) Time-intensity: the temporal aspects of sensory perception. Food Technol 40(71–78):82

Peyvieux C, Dijksterhuis G (2001) Training a sensory panel for TI: a case study. Food Qual Prefer 12:19–28

8.3   For Instructors and Assistants

8.3.1   Notes and Keys to Successful Execution

1. If the class does not have access to a computer data capture system such as Compusense, SIMS, or FIZZ, that part of the lab may be omitted. Different systems may provide different forms of parameter output (Imax, DUR, etc.) and may provide either individual or group information. The exercise may need to be tailored to adjust for the information provided by your system. If you have a limited number of tasting booths or stations, the class may have to be further divided or given time appointments to come to the sensory facility.

2. Students may calculate their own Tmax, Imax, DUR, and AUC values in class for their paper ballots. As noted, the AUC can be estimated geometrically, using triangular approximation or by cutting out the resulting curve boundaries and weighing the paper. Due to the uniform density of most papers, this method works quite well.

3. The hot sauce may simply be refrigerated. TAs should record the temperature at which it is served. A suggested maximum for the chilled version is 10 C. If several small bottles of hot sauce are used, they may be kept in an ice bath. It may be possible to obtain individual size servings or foodservice packets of some sauces. This will facilitate students all having the same starting temperature rather than passing a single bottle from person to person wherein warming can occur.

4. Samples may be labeled with three-digit codes or simply "RT" and "chilled." The study is obviously not blind with regard to temperature. Samples may be presented in cups, and if so, the tasting method (dipping a plastic spoon into the sample) should be illustrated. Alternatively, students may be provided with their own small bottles of sample and place the three drops on the spoon as instructed.

5. The sample ballot provides a line scale, but a category ballot is equally appropriate. Using a category scale may provide more of a contrast with computer systems as most of these will use a vertical line scale or "thermometer" type of display for the TI capture.

6. If the scales are in a comparable range, or converted to a common scale such as a 100-point maximum basis, a two-way ANOVA may be done to compare both temperatures and method (stopwatch vs. computer).

7. Time intervals are suggested on the sample ballots. These can be adjusted. A common problem is that the intensity is still nonzero when the last time sample is reached. This can provide an opportunity for discussing censored or truncated data and some of the issues involved.

8.3.2   Equipment

Computerized data capture system, stopwatch, scissors, and a balance or scale for weighing paper, if AUC is estimated by the cutout method

8.3.3   Supplies

Cinnamon-flavored gum, hot sauce (Tabasco or similar), serving cups, rinse cups, spit cups, crackers, water, napkins, spill control, trash disposal items. Trays are useful but not necessary because there are only two samples (per method) and the gum.

8.3.4   Procedure

There is no special sample preparation needed for this exercise other than those discussed above in the Notes section. For the paper ballot method, the instructor or teaching assistant should control the stopwatch and issue verbal cues ("rate now") to the class. Do not allow students to time themselves by looking at a clock or wristwatch.

8.4   Appendix: Sample Ballots and Data Sheets

Time Intensity Score Sheet

Place the *gum* in your mouth and begin to chew.

IMMEDIATELY score the first rating for **cinnamon flavor intensity** at the 0 sec interval by placing a horizontal mark anywhere along the line.

Continue to chew the sample and rate the **cinnamon intensity** every 20 seconds for 2 minutes.

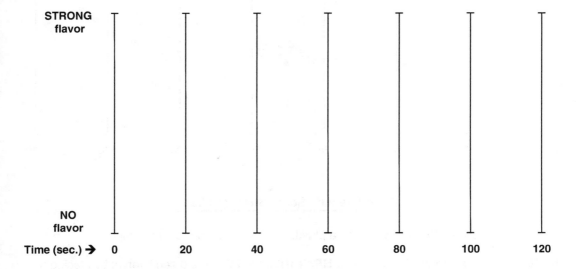

Time Intensity Score Sheet (room temperature)

Place three drops of *Room Temperature sauce* on a plastic spoon. Taste the entire sample.

IMMEDIATELY score the first rating for **HEAT INTENSITY** at the 0 sec interval by placing a horizontal mark anywhere along the line.

Continue to chew the sample and rate the **HEAT INTENSITY** every 30 seconds for 3 minutes.

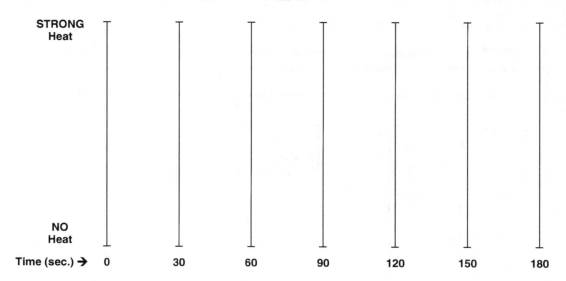

STRONG Heat						
NO Heat						
Time (sec.) ➔ 0	30	60	90	120	150	180

Time Intensity Score Sheet (chilled)

Place three drops of *Chilled sauce* on a plastic spoon. Taste the entire sample.

IMMEDIATELY score the first rating for **HEAT INTENSITY** at the 0 sec interval by placing a horizontal mark anywhere along the line.

Continue to chew the sample and rate the **HEAT INTENSITY** every 30 seconds for 3 minutes.

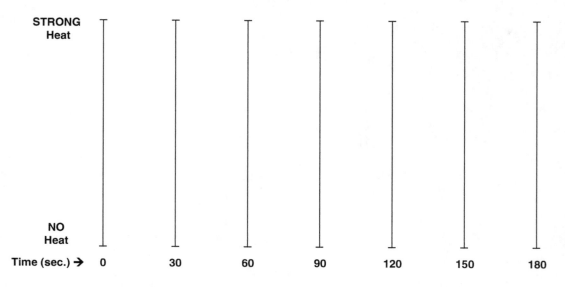

STRONG Heat						
NO Heat						
Time (sec.) ➔ 0	30	60	90	120	150	180

Sample 1: Time intensity raw data sheet (reproduce as necessary)

ID	0	30	60	90	120	150	180	Imax	Tmax	DUR	AUC
Mean											
SD											
SE											

Sample 2: Time Intensity Parameter Summary Sheet (reproduce as necessary)

Name	Room Temp.				Chilled			
	I_{max} (cm)	T_{max} (sec)	DUR (sec)	AUC (or wt. in grams)	I_{max} (cm)	T_{max} (sec)	DUR (sec)	AUC (or wt. in grams)
Mean								
SD								

Flavor Profile Method

9.1 Instructions for Students

9.1.1 Objectives

To introduce descriptive panel approaches to food analysis.

To participate in an analytical sensory process.

To appreciate the value of reference materials in descriptive analysis.

To gain experience working in a small group.

9.1.2 Background

The flavor profile method originated with the Arthur D. Little consulting group in the late 1940s (Caul, 1957). At the time it provided a general tool for characterizing the flavors of complex food products and replaced the approach of using only one expert taster with the use of a panel of highly trained individuals with general experience in flavor sensory analysis. The method proved valuable for examining flavor differences among foods that were the results of ingredient and/or processing changes (Keane, 1992).

Unlike more modern practices, the traditional flavor profile uses a consensus procedure. The group of trained individuals decides through discussion how the food is to be scored on various attributes (also chosen by the group). Individual judges begin by tasting and working independently from one another but come together as a group to discuss the attributes and finish the

flavor profile. In descriptive analysis, statistical analysis is generally preferred over consensus-based procedures, but the flavor profile method continues to be used by some laboratories (Lawless and Heymann, 2010).

In the exercise that follows, we will generate flavor profiles of popular commercial blended vegetable juice product by working in small groups with both references and with the finished product. The reference materials are the isolated ingredients of the product. They are tasted first, in order to provide a short training experience and then used to identify the flavor components that are most pronounced in the juice. This exercise will also demonstrate how a complex finished food product can be analytically fractionated by a judge into its component tastes and flavors.

9.1.3 Materials and Procedures

9.1.3.1 Materials (for Each Group of 4–5 Students)

You will be given samples of pureed or juiced components of an eight-component vegetable juice including tomato juice, pureed beets (fresh and cooked), pureed carrots (fresh and cooked), pureed celery (fresh), pureed parsley (fresh), pureed romaine lettuce (fresh), pureed spinach (fresh and cooked), pureed watercress (fresh), and salt. You will also be provided with the finished product. These samples will be shared among the group members who will take individual samples from the group containers. In addition to the ingredients, obtain enough small cups and plastic spoons for each item and for each person in your group, as well as water, spit cups, napkins, and crackers.

9.1.3.2 Procedures

1. Assemble groups of four or five students each. Elect a group spokesperson (panel leader).
2. Take a personal sample of about 30 ml from each of the large cups of reference standards and place it in a smaller cup as your personal training sample. Mark the identity of the sample with a pen or odorless marker.
3. When you have assembled your personal reference set, taste each of the components of the juice listed above. Taste only the components and not the finished product until you are instructed to do so. The components may be sampled in any order. This is done to train your senses to recognize the individual flavor of each item. Pay close attention to the flavors present in each of the components, making notes as you go if you wish.
4. When everyone in the group has sampled all of the components, you may begin your evaluation of the finished product and then construct your group flavor profile. Elect one of your group to be the panel leader who will record and report your completed profile. Begin by tasting the finished product. Continuing to work individually, write down all of the flavors you experience in the product and the order they appear to you. You may use a blank sheet of paper or a ballot, if provided, for this part of the exercise. Use the following rating scale to denote the intensity of each perception: 0=not present,)(= threshold, just barely perceptible, 1=weak, 2=moderate, and 3=strong. You may use half points on the scale if you feel it's necessary.
5. When everyone in your group has completed her or his individual profile and is satisfied with it, proceed to the group-consensus step. The panel leader will lead the discussion, submit the group's profile, and serve as a spokesperson. Through group discussion compare the flavor notes, along with their intensities. Try to come to agreement (i.e., consensus)

among your group on the intensity of each flavor note. Do not average the ratings from your group members (that would be applying a statistic!). Once your group has reached this consensus, record your flavor profile on the group ballot and give it to a TA for tabulation. Your instructor may ask for your profile verbally and show it to the class to compare each group's results.

9.1.4 Data Analysis

There is no data analysis for this exercise.

9.1.5 Report

This lab report may not follow the standard format unless you are instructed to do so on your syllabus, lab manual, or course website. Obtain the data table containing all of the group flavor profiles generated in the above exercise.

Provide a short essay to address the questions below, in complete English sentences. Your grade will reflect the carefulness, depth, and quality of your discussion. Refer to the chapter on descriptive analysis (Chap. 10) in Lawless and Heymann, 2010 for additional ideas or use the papers cited below. You may append the group data table, to illustrate your conclusions.

Discuss the level of agreement among the different groups and within your group by addressing these issues. Do not provide numbered answers but construct a coherent essay in English prose:

1. Generally did the groups find the same major flavor notes in the juice?
   (a) Did the groups tend to agree about the relative intensities of the flavor notes?
   (b) What flavors were consistently not present?
   (c) What kind of pattern would you expect if responding was random and there was no agreement?
2. What did you notice about the stronger versus the middle or weaker flavor notes in terms of between-group agreement?

3. What, if any, problems did you encounter with the consensus procedure?
   (a) Was it easy or difficult to reach agreement within your group about the intensities of specific flavor notes? Why or why not?
4. Do you feel that group discussion yields as accurate description of the product or would you prefer a statistical approach (e.g., averaging individual ratings)? Why or why not?
5. Were there any flavor notes in the juice that you felt you did not have a reference for (additional tastes or flavors that were not in the reference set)?

9.2   For Further Reading

Caul JF (1957) The profile method of flavor analysis. Adv Food Res 7:1–40
Keane P (1992) The flavor profile. In: Hootman RC (ed) ASTM manual series MNL 13 manual on descriptive analysis testing for sensory evaluation. American Society for Testing and Materials, West Conshohocken, PA, pp 5–14
Lawless HT, Heymann H (2010) Sensory evaluation of foods, principles and practices, 2nd ed., Springer Science+Business, New York

9.3   For Instructors and Assistants

9.3.1   Notes and Keys to Successful Execution

1. General flow. There are two phases to this lab, a training phase and a test phase. In the training phase, each student will spoon out his or her individual training samples into small cups that they have labeled with the identity of the sample using an odorless marker. They should take at least a half hour to taste all the references and make any notes they wish on a piece of scratch paper. After they have tasted everything in the training phase, they should elect a group leader and proceed to do their ratings. A blank sheet should be used by each student for recording his/her individual flavor ratings.

2. Groups may not come to consensus by averaging. This is tempting but the class needs to be monitored to insure they are having a discussion. Averaging is not allowed. The group leader should report the consensus profile.

3. Record the consensus profiles in front of the class when all groups are finished. A blackboard, whiteboard, or an overhead may be used to display the profiles and to discuss the degree of agreement among the groups. This is easiest to see if the group lists the flavors in descending order—strongest to weakest and omits those not present. Remind them that those flavor terms "not present" constitute another form of between-group agreement. Usually, there is strong agreement on the most intense flavors and flavors not present, with less agreement in the middle. However, the top three or four are usually the same. You can circle them for emphasis.

4. Preparation. Allow sufficient time for shopping and kitchen processing. TAs should not underestimate the time required for this lab. Some materials may be difficult to find, such as watercress. Multiple stores may be visited or their websites inspected before shopping to insure they have everything (or contact the produce manager).

5. The lab groups and samples can be quite colorful when all the samples are spread out, so this is a photo opportunity if you wish to record your class. Permission for photos (from people being photographed) may be necessary in some locations.

6. It is recommended to steam the vegetables marked "cooked." Alternatively they may be microwaved in a glass container covered with plastic wrap or boiled (but boiling removes some flavors).

7. A reduced sodium version of the finished product may be used in place of the traditional version of the juice product. If so, it is advisable to include a reference standard of potassium chloride or a salt substitute high in potassium to illustrate its taste properties. Some students may be highly sensitive to the bitter or metallic tastes from potassium, others

are not. This can be a discussion point about sodium-reduced products.

8. Provide a common spoon in each group sample to dole out the individual practice samples. Do not allow students to use their individual spoons to take their individual portions from the group container—this is microbiologically unsafe as they may have used their individual spoons for tasting already.

9. It is useful to have an overhead of the intensity category scale (zero to three, with "threshold" symbols), or you can write it on a blackboard in front of the class. Allow half points. This can be a discussion point concerning scale types.

9.3.2 Equipment

Food processors or blenders, two or more recommended. Spatulas. Pots/pans for vegetable steaming or boiling. Conventional stove top/range or microwave oven. Colander/strainer

9.3.3 Supplies

10–12 1 ounce (30 ml) or larger cups per student for individual samples

10–12 8 ounce (250 ml) or larger cups for each group for serving reference samples and tasting the V-8 in the test phase

Odorless markers, one per student or per pair of students

Napkins, rinse cups, spit cups, unsalted crackers, and rinse water

Items needed for spill control and trash collection

9.3.4 Food Samples Required

Tomato juice (unsalted if possible, but beware of low sodium items which may contain large amounts of potassium chloride, see Note 7 above)

Pureed beets (fresh and cooked)
Pureed carrots (fresh and cooked)
Pureed celery (fresh)
Pureed parsley (fresh)
Pureed romaine lettuce (fresh)
Pureed spinach (fresh and cooked)
Pureed watercress (fresh)
Table salt (Kosher recommended)
V-8® Juice (Campbell's)

9.3.5 Procedures

9.3.5.1 Preparation

Samples marked "cooked" should be steamed or boiled in a small amount of water until soft. Each sample should be pureed in a food processor or blender until a smooth consistency is obtained.

If samples are prepared the day before, they should be refrigerated in closed containers in order to preserve the flavor. They should be brought up to room temperature at least an hour before class.

Be sure to have sufficient samples for each student and each lab group. Six to eight ounces (240 ml) is a minimum for a lab group of 4–5 students.

9.3.5.2 Classroom Procedure

Be sure to allow sufficient time for the two phases (training and evaluation) and emphasize the level of concentration needed in the training phase. Discussion is permitted. Once the format evaluation commences, students should work as if they are in isolated booths (although they are not). Consensus discussion only starts when all members of the group have completed their blank-sheet profiles. Do not allow averaging of data to achieve consensus.

9.4   Appendix: Sample Key and Class Data Sheet

Flavor Profile

Sample Legend (use if necessary, or simply label cups with the contents)

Code	Sample Content
100	V-8 Juice
200	Pureed Beets (Fresh)
201	Pureed Beets (Cooked)
300	Tomato Juice
400	Pureed Carrots (Fresh)
401	Pureed Carrots (Cooked)
500	Pureed Celery (Fresh)
600	Pureed Parsley (Fresh)
700	Pureed Romaine Lettuce (Fresh)
800	Pureed Spinach (Fresh)
801	Pureed Spinach (Cooked)
900	Pureed Watercress (Fresh)
1-oz. Cup	Table Salt

Flavor Profile Class Data Sheet (Blank)
List flavors from strongest to weakest. Omit zeros (flavors not present)

Flavor	Group 1	Group 2	Group 3	Group 4	Group 5	Group 6	Group 7	Group 8

Rating Scale:
0 = not present   )(= threshold, just barely perceptible   1 = weak   2 = moderate   3 = strong

Lab 7– Flavor Profile Optional Class Data Sheet (with descriptors)

Flavor	Group 1	Group 2	Group 3	Group 4	Group 5	Group 6	Group 7	Group 8
Beets (Fresh)								
Beets (Cooked)								
Tomato								
Carrots (Fresh)								
Carrots (Cooked)								
Celery (Fresh)								
Parsley (Fresh)								
Romaine Lettuce (Fresh)								
Spinach (Fresh)								
Spinach (Cooked)								
Watercress (Fresh)								
Salt								

Rating Scale:
0 = not present   )(= threshold, just barely perceptible   1 = weak   2 = moderate   3 = strong

Introduction to Descriptive Analysis

10.1 Instructions for Students

10.1.1 Objectives

To become familiar with methods used for term generation.

To become familiar with the criteria for including terms in a descriptive ballot.

To become familiar with descriptive analysis data collection and analysis.

10.1.2 Background

In most descriptive analysis procedures, a critical step is the choice of terms to be included on the descriptive scorecard or ballot. This is accomplished through a creative process similar to group brainstorming in which all possible words that might be used to describe the product or product category are collected from a panel of respondents. A panel leader will often provide some focus in this step by imposing some general categories of description on the product such as appearance, aroma, flavor, texture, and residuals.

Once the potential term list has been assembled, the number of terms is reduced by eliminating redundant or overlapping terms, those terms that are vague (e.g., refreshing) and those terms that have an affective meaning (e.g., distasteful). In addition, terms that have complex meanings are broken down into simpler components if possible (e.g., creamy). The term list continues to be narrowed, perhaps over several training sessions, until the panel agrees on the meaning of each term and that the list of terms adequately describes the product or the product category of interest. If possible, reference standards are found to illustrate the sensory properties associated with each term.

Finally, the panel must choose anchor terms for the high and low ends of the scales on the ballot. These anchor terms, and physical references if necessary, help the panel to rate similar sensations in similar regions of the scales. In the exercise that follows, the instructor will act as panel leader, and the class will act as the panel. A descriptive ballot will be generated from a set of fruit juice products and later be used to rate another small set of products.

In the descriptive analysis method, panelists, working individually, quantitatively specify the perceived intensities of a group of attributes that is specific to a particular product or class of products. The descriptive attributes are chosen and refined by the panel in the ballot generation step of descriptive analysis. A panel of trained individuals may be given several practice sessions with the product or the product category of interest to make sure that each member of the panel is using the attributes and scales in the same fashion. The inclusion of practice trials can eliminate errors such as misunderstanding term definitions, anchor terms, or mentally reversing the scale so that these points of variation are weeded out of the data when the actual test products are presented.

H.T. Lawless, *Laboratory Exercises for Sensory Evaluation*, Food Science Text Series 2,
DOI 10.1007/978-1-4614-5713-8_10, © Springer Science+Business Media New York 2013

The intensity ratings are usually made on category or line scales. Data are analyzed using analysis of variance (ANOVA). If there are only two products, *t*-tests are appropriate. Descriptive statistics, including means and standard deviations and standard errors (of the mean), are calculated for each attribute and product. Planned comparisons are then performed to determine differences between means, when there is a significant product effect in the ANOVA. Some common tests include Duncan's multiple-range statistic, Tukey's honestly significant difference (HSD), and the least significant differences (LSD).

Once descriptive statistics have been calculated and compared, the data are plotted so that the relationships between means and attributes may be visually compared. A frequently used graph for descriptive analysis is the "spider" plot or radial-axis graph that shows the means of several products on axes that radiate from a central point in the graph. Each axis (or radius or "spoke") on such plots represents a single attribute in the descriptive analysis. If five to eight attributes are included on each plot, the mean ratings given a single product will be represented by a simple polygon in the graph. Products may then be visually compared by examining the individual shapes of two or more polygons on a particular graph. Examples of spider plots may be posted on your class website. Additional information about descriptive analysis technique, data analysis, and graphing is found in Lawless and Heymann, Chap. 10 and the papers cited in the reference list.

This lab has two phases. In the first, the terms are generated, and the ballot is then constructed. In the second phase of the lab, you will rate three juice products, using the descriptive ballot generated in the first part of this exercise.

10.1.3 Materials and Procedures, Part I: Term Generation/Ballot Construction

10.1.3.1 Materials
Three samples of grape juice or apple juice with diverse sensory characteristics should be purchased. These may range from a non-clarified organic 100 % juice to a fully clarified partial-juice product. A similar juice product may be substituted depending upon availability.

10.1.3.2 Procedures
The first product will be tasted, and students will write on a blank sheet of paper all sensory properties that they individually perceive. Divide your blank sheet into five categories of sensations, including appearance, aroma, flavor/taste, mouth-feeling factors, and residual sensations. These should be evaluated in that order, writing down all the aspects of the product that you notice. "Residuals" include all sensations remaining after expectorating the sample. Please avoid terms that refer to hedonics ("good," "unacceptable"), terms that are complex combinations of several factors ("refreshing") or vague and not specific ("natural").

The instructor will then collect all of the terms from the class in a group discussion. Group discussion may be used at this point to eliminate redundant, vague, or affective terms.

Taste the second product, and students and instructor will repeat the above exercise. Your term list may be expanded upon or narrowed further with the information gathered from a second product. In addition, by sampling a range of products, the panel can begin to get an idea of what anchor terms should be used for each scale and whether or not physical reference materials are necessary to clarify the meaning of specific terms.

The tasting and term discussion may be repeated using a third product to further refine the term list on the ballot. The ultimate aim of the exercise is to generate a ballot for use in the second half of this lab.

10.1.4 Materials and Procedures, Part II: Descriptive Evaluation

(Note: There will generally be a half-hour break in between part I and part II to allow the TAs to construct the ballots.)

10.1.4.1 Materials
Three samples of grape or apple juice.

10.1.4.2 Procedures

The TAs, staff, or instructors will provide you with three samples of apple juice, each identified with a three-digit blinding code. Evaluate and rate the products using the ballot provided.

10.1.5 Data Analysis

Check with your instructor to see if you are required to do ANOVAs (this will usually be divided up among the class) and LSD tests or whether results will be provided to you for interpretation and graphing. It is recommended that each student be assigned one attribute to analysis using two-way ANOVA with products and panelists as factors. See the ANOVA appendix in Lawless and Heymann (2010) for examples. Note: If there are missing data from any student on any single attribute, that student must be dropped from the analysis of that scale.

10.1.6 Reporting

Write up the results in the standard lab format unless instructed otherwise. In the results, name the attributes that showed significant differences, and tell how the products were different (not just significance). For example, "product 387 was sweeter, less sour, and less astringent than product 582." Discuss only the significant differences. Do not waste time on nonsignificant differences.

Make a bar graph or spider plot of the mean ratings to show the differences among the products.

Place a table of all the mean values and standard errors of the mean for all attributes in an appendix. List products in columns and attributes in rows. Show differences in means in rows by LSD tests by using different letters. If means are not significantly different, give them the same letter. Means not sharing the same letter are significantly different. For example, if 387 is significantly sweeter than 582, they could have the letters A and B. If neither is significantly different from product 245, which is intermediate, it gets the letters A and B to indicate its statistical overlap with both 387 and 582.

10.2 For Further Reading

An simple example of descriptive analysis in practice can be found in: Lawless HT, Torres V, Figueroa E (1993) Sensory evaluation of hearts of palm. J Food Sci 58:134–137

Lawless HT, Heymann H (2010) Sensory evaluation of foods, principles and practices, 2nd ed., Springer Science+Business, New York

Meilgaard M, Civille GV, Carr BT (2006) Sensory evaluation techniques, 4th edn. CRC Press, Boca Raton, FL

Stone H, Sidel J, Oliver S, Woolsey A, Singleton RC (1974) Sensory evaluation by quantitative descriptive analysis. Food Technol 28: 24–29, 32, 34

10.3 For Instructors and Assistants

10.3.1 Notes and Keys to Successful Execution

1. General flow. There are two phases to this lab, a term generation phase and a testing phase.

2. The three samples used in the term generation phase should be diverse in sensory characteristics. This may take some preliminary work and benchtop tasting of the staff to decide on the best products to include. A locally manufactured small-scale organic product will sometimes exhibit different characteristics than a mass-produced version. Also, a processed juice product such as a children's juice drink made from concentrate or a combination of different fruit sources can be substituted for variation. The test samples should contain at least one new product that was not seen previously in the preparation phase, if possible.

3. There is usually a necessity of a break in between the ballot generation and test phase to allow the TAs or staff to compile the ballot, unless one has been prepared ahead of time or from a previous year. Students may be recruited to help pour samples and/or label cups while the ballot is being compiled and photocopied.

4. Being a panel leader in a term generation exercise is a challenging task and should not be undertaken without experience, guidance, and/or practice. Guidelines are listed at the end of this section.
5. Analysis options. TAs can perform the analysis of variance and send the results, means, and standard errors to the students. Alternatively, you can ask each student to do one two-way ANOVA (on a single attribute) and LSD tests, and post or e-mail the results to the TA for compilation. If the class is large enough, assign each attribute to two different students so you can check their work. This may require an extra week between execution and reporting, unless you press them for faster turnaround.
6. It is recommended that you use a product you have some technical knowledge about. If your department is strong in dairy, for example, you could use a cheese or yogurt product instead of fruit juice.

10.3.2 Equipment

None. A blackboard, white board, or flip chart is needed for term generation. TAs will need access to a photocopy machine to print ballots during the break, unless a computer-assisted system for data collection is used.

10.3.3 Supplies

6–8 1 oz (30 ml) or larger cups (100 ml recommended) per student for individual samples
2–8 oz (250 ml) or larger cups for rinse and expectoration
Other: napkins, rinse cups, spit cups, unsalted crackers, and rinse water
Items needed for spill control and trash collection
Samples required: three or four diverse tasting samples of commercially available fruit juice of the same type (apple or grape, not both). A minimum of 30 ml is required; 50–100 ml per student in phase I is

recommended. Approximately 2 l of juice will provide two 50-ml samples per student.

10.3.4 Preparation

Samples may be poured ahead of time and served at room temperature. All samples should be labeled with random three-digit codes. Paper labels on cups are recommended. If a marker is used, it should be odorless.

10.3.5 A Quick Look at Panel Leadership in Term Generation and Ballot Development

1. Stand in front of the class. When they are done tasting the first sample, you can write down the terms they found.
2. Use the five categories of appearance, aroma, flavor/taste, mouthfeel, and residual as headings on a blackboard to collect terms.
3. On the first pass, do not judge terms but simply write down everything that is generated by the class. Ask them to call out their terms verbally. Do not judge them or indicate what is right or wrong.
4. On the second product, you can begin to eliminate hedonic terms. You can also try to eliminate redundant terms after some discussion: Is sour the same as acidic (probably yes)? Astringency vs. drying?
5. For terms that are vague or combinations, try to get them to clarify. What do you mean by "refreshing?" What does that feel like?
6. Stress that if a term is only generated by one or a few people, in a real DA, they would be asked to go and find reference standards to illustrate what they mean by the term, to be evaluated in a future training session. But "for now," we will not have time to do that.
7. Don't forget to get anchor words (none–extremely strong, light–dark, thin–thick) for each scale.
8. Have the TAs or staff take only the frequently used and consensus words to construct the ballot. 15–20 terms should be sufficient.

10.4   Appendix

10.4.1 Descriptive Term Ballot

DESCRIPTIVE TERMINOLOGY EXERCISE

1. Observe all the visual attributes of the product and list under APPEARANCE.
2. Inspect the aroma by sniffing the headspace and list all olfactory attributes under AROMA.
3. Next, taste the product and note all taste and flavor attributes under TASTE/FLAVOR.
4. Observe all mouthfeeling attributes and tactile properties and list under MOUTHFEEL.
5. Expectorate the product and note all residual characteristics under RESIDUAL.

Appearance	Aroma	Taste/Flavor	Mouthfeel	Residual

In subsequent samples, you may circle attributes that have already been noted.
You may circle them more than once.

10.4.2 Sample Apple Juice Ballot

Sample apple juice ballot
(You should fill in the terms as you develop them in the exercise)

Appearance

1. Clarity

Very clear   Very Cloudy
❏   ❏   ❏   ❏   ❏   ❏   ❏   ❏   ❏   ❏

2. _____

❏   ❏   ❏   ❏   ❏   ❏   ❏   ❏   ❏   ❏

3. _____

❏   ❏   ❏   ❏   ❏   ❏   ❏   ❏   ❏   ❏

4. _____

❏   ❏   ❏   ❏   ❏   ❏   ❏   ❏   ❏   ❏

Aroma
5. _____

❏   ❏   ❏   ❏   ❏   ❏   ❏   ❏   ❏   ❏

6. _____

❏   ❏   ❏   ❏   ❏   ❏   ❏   ❏   ❏   ❏

7. _____

❏   ❏   ❏   ❏   ❏   ❏   ❏   ❏   ❏   ❏

8. _____

❏   ❏   ❏   ❏   ❏   ❏   ❏   ❏   ❏   ❏

9. _____

❏   ❏   ❏   ❏   ❏   ❏   ❏   ❏   ❏   ❏

Taste/Flavor

10. _____

❏   ❏   ❏   ❏   ❏   ❏   ❏   ❏   ❏   ❏

11. _____

❏   ❏   ❏   ❏   ❏   ❏   ❏   ❏   ❏   ❏

12. _____

❏   ❏   ❏   ❏   ❏   ❏   ❏   ❏   ❏   ❏

13. _____

❏   ❏   ❏   ❏   ❏   ❏   ❏   ❏   ❏   ❏

14. _____

❏   ❏   ❏   ❏   ❏   ❏   ❏   ❏   ❏   ❏

15. _____

❏   ❏   ❏   ❏   ❏   ❏   ❏   ❏   ❏   ❏

Mouthfeel
16. _____

❏   ❏   ❏   ❏   ❏   ❏   ❏   ❏   ❏   ❏

17. _____

❏   ❏   ❏   ❏   ❏   ❏   ❏   ❏   ❏   ❏

18. _____

❏   ❏   ❏   ❏   ❏   ❏   ❏   ❏   ❏   ❏

Residual
19. _____

❏   ❏   ❏   ❏   ❏   ❏   ❏   ❏   ❏   ❏

20. _____

❏   ❏   ❏   ❏   ❏   ❏   ❏   ❏   ❏   ❏

Use of Reference Standard in Panel Training

11

11.1 Instructions for Students

11.1.1 Objectives

To familiarize students with the use of reference standards in panel training for intensity and quality judgments.

To show how calibrated scales can be used in descriptive analysis and related methods.

To show how vocabulary development can be enhanced with use of specific physical examples.

11.1.2 Background

Trained panels are used in sensory evaluation in several specific scenarios. Probably the most common is in descriptive analysis, in which a panel must develop both a vocabulary of terms to describe the products as well as a frame of reference for what constitutes examples of high and low amounts of each attribute. Another common example is in quality control, where products may be evaluated on key attributes or examined for defects. Defect or taint analysis has a long history in the grading of standard commodities such as fluid milk and some common kinds of cheese. An important aspect of panel training is the use of reference standards to train and exemplify these specific sensory aspects. Reference standards are also used in some sensory methods to exemplify specific intensity levels of a given

attribute. A good example is the pepper heat scale developed by ASTM in its various standards for evaluating the heat intensity of various pepper products and derivatives.

The use of reference standards for specific terms is useful in several ways. Not everyone will have the same idea of what is meant by a "green" aroma, for example, when they enter panel training. Using some reference standards, often with specific chemicals such as cis-3-hexenol, can give everyone on the panel roughly the same idea of what you are talking about. Some reference standards are everyday products or treatments of common products, such as using the juice from canned asparagus to add to a wine to show a green or asparagus-type aroma in that specific product. This aids in what is known as concept formation or concept alignment. The idea is to get all the panelists thinking about the same experience when they try to find that attribute in a product and rate its intensity. The overall goal is to lower the interindividual variability by providing a common reference or benchmark for the entire panel. Good examples of reference standards for flavor and aroma terms are found in the articles on the wine aroma "wheel," and various similar classification systems have been developed for other commodities. Classic examples of product defect judging can be found in the dairy literature, with recipes for illustrating specific defective characteristics.

Sometimes it is also useful to calibrate panelists with intensity scales. Important historical examples can be found in the literature on the

H.T. Lawless, *Laboratory Exercises for Sensory Evaluation*, Food Science Text Series 2, DOI 10.1007/978-1-4614-5713-8_11, © Springer Science+Business Media New York 2013

texture profile method. In texture profile training, panelists are given examples of different levels of hardness in order to calibrate themselves regarding the force necessary to bite through a food sample. This scale used common easily obtainable products as the reference materials for different levels of hardness. Other scales for other texture attributes were also part of that method. In some descriptive analysis techniques, notable the Spectrum™ method, a universal 15-point category scale is used with reference items for the intensity of different tastes, aromas, and flavors. In this lab, we will look at an example using a 15-point scale for sweetness.

Note that these ideas about training, calibration, and uniformity assume that people are having similar experiences from the standards, which is not necessarily so. Even if their experiences are different, then, the idea is that they can translate or shift them to conform to the group judgment. This is something of a violation of the psychophysical model. In a psychophysical method, the subject's only job is to report on his or her sensations (their strength or intensity) not to translate into someone else's arbitrary scale. So in spite of the overall utility of this general approach, the student should realize that not all people's experiences are bound to be the same, even from the same stimulus. It is notably difficult, for example, to get people to agree on the same bitterness level to a given chemical or flavor material. There is simply too much genetic variation. For that reason, some scales have been anchored to more general experiences in one's memory, an ongoing debate discussed in Chap. 7 in Lawless and Heymann.

Further information on reference standards can be found in articles by Rainey (1986), the texture profile paper by Szczesniak et al. (1963), the ASTM example of pepper heat (ASTM 2008), and papers on the wine aroma wheel and its standards (Noble et al. 1987). Recipes for dairy product defects are throughout the text by Bodyfelt et al. (1988) or in the updated edition (Clark et al. 2009) on pages 551–560. The general use of intensity references for descriptive scales for taste and flavor is in Meilgaard et al. (2006). A good example of a lexicon for a specific aroma attribute, green flavors, can be found in Hongsoongnern and Chambers (2008).

This lab is divided into four exercises or "options." Your instructor may wish to do all four exercises or just some of them, so check your course syllabus or website for further details. Each option has a training phase and a test phase.

11.1.3 Materials and Procedures

11.1.3.1 Materials

Option 1: Reference standards for sweetness intensity. In the training phase you will be given four references representing scale points as follows: 2, 5, 10, and 15 on a 15-point sweetness scale, consisting of various concentrations of sucrose in water. In the test phase you will be given various examples of a fruit beverage to rate for sweetness intensity, using the scale you have just practiced.

Option 2: Reference standards for hardness intensity. You will be given nine references meant to represent levels of hardness on a nine-point scale.

Option 3: Reference standards for wine aroma. You will be given five or six references as wine samples in covered tasting glasses to smell. Each will be labeled with an aroma characteristic. There will be another set of five or six blind-coded references for you to try and smell and label with the correct aroma characteristic, based upon your training.

Option 4: Dairy product defects in fluid milk. You will be given four or five samples of fluid milk to taste with different flavor characteristics that have been branded as defects in dairy product judging. After you are done, you will be given four or five blind-coded samples to taste.

11.1.3.2 Procedures

Option 1: Reference standards for sweetness intensity. Taste the references in the order from lowest to highest. Be sure and rinse your mouth between each sample. Once you have tasted all of them, you can "test" yourself by using the four blind-coded samples on your tray. Try to assign each blind-coded sample to one of the four reference levels. The TA or instructor will provide a code sheet to show you the "correct" answers. If you were incorrect on any levels, go back and try the reference samples one more time, working again from low to high.

When you are done training yourself, try the three blind-coded beverages and assign a sweetness rating based upon your understanding of the scale.

Option 2: Reference standards for hardness intensity. Take each sample and place it between your molar teeth, and bite through it once noting the force required. If you are unable to bite through the hardest two samples, then omit those samples. When you are done, you will be given three test samples to bite through and rate for hardness, using the nine-point scale you have learned.

Option 3: Reference standards for wine aroma. You will be presented with an array of wine glasses with watch glass or similar coverings. Swirl the sample gently and then remove the cover and quickly smell the headspace. Note the name of the aroma attribute listed on the glass. You may repeat smelling it once again if that sensation was not clear to you. When you have sampled all the glasses, move on to the tray of test samples. Sniff the headspace in each one, and choose one of the aroma terms you learned from the reference samples, by filling in the random three-digit code on the checklist of terms. If you feel a more than one sample has that characteristic, you may enter two code numbers for that term. If you feel a sample has two characteristics, you may note both of them.

Option 4: Reference standards for dairy defects in fluid milk. You will be presented with an array of glasses or plastic cups, labeled with the name of a specific dairy product defect characteristic. Taste each sample and spit it out, being careful to note the aroma characteristic of that sample. You will also be given a large sample of fresh milk with no defect for comparison. You may wish to rinse with water between each sample to prevent flavors building up or carrying over into the next sample, if you find that helpful. When you are done inspecting the reference standards, you will be given a tray with four or five milk samples in

covered containers. Try to identify the defect, based upon your training, and list it next to the three-digit code. If you feel you can identify the defect by simply smelling, you do not need to taste the sample. If you feel the sample has more than one defect, record both. If you feel the milk has no defect, leave it blank.

When you have completed each exercise, record your choices/responses on the master sheets.

11.1.4 Data Analysis

The TAs or instructor will provide you with a spreadsheet of the data for the intensity references. A separate sheet will show the term choices for the aromas and dairy defects, by frequency counts.
1. For the intensity references, compute means, standard deviations, and standard errors of the mean. Place these into a simple table, one for the sweetness and one for hardness if you performed both options.
2. Perform a simple one-way ANOVA on the samples in each option. Were there significant differences?
3. For the term frequencies, make a simple bar graph for each sample and the choices people made.

11.1.5 Reporting

This lab will use the standard format unless your instructor advises otherwise.

In reporting your results, do the following:
1. Refer to your tables and graphs and describe what happened.
   (a) For the intensity references, were some products *more variable* than others? If so, why do you think this could have happened?
   (b) Were there significant differences based upon your ANOVA's? (Describe.)
   (c) For the aroma and defect references, were there some characteristics that seemed more difficult than others, as shown by errors? Why do you think that could have happened?

In discussion, address the following issues:

2. Did you feel you could perform these tasks better due to the "training?"
   (a) What do you think the data would look like if there were no training phase?
3. How much training do you think would be necessary (how many sessions) in order to get to perfect (or near perfect) performance?
4. How would you monitor a panelist's progress as training continued?

11.2 For Further Reading

ASTM (2008) Standard test method for sensory evaluation of red pepper heat. Designation E 1083–00. In: Annual book of ASTM standards, vol 15.08, End use products. American Society for Testing and Materials, Conshohocken, PA, pp 49–53

Bodyfelt FW, Tobias J, Trout GM (1988) Sensory evaluation of dairy products. Van Nostrand/ AVI Publishing, New York

Clark S, Costello M, Drake M, Bodyfelt F (eds) (2009) The sensory evaluation of dairy products. Springer Science + Business, New York. See Appendix F, Preparation of samples for instructing students and staff in diary product evaluation by Costello M, Drake M, pp 551–560

Hongsoongnern P, Chambers EC IV (2008) A lexicon for green odor and flavor characteristics of chemicals associated with green. J Sens Stud 23:205–221

Meilgaard M, Civille GV, Carr BT (2006) Sensory evaluation techniques, 4th edn. CRC Press, Boca Raton, FL

Noble AC, Arnold RA, Buechsenstein J, Leach EJ, Schmidt JO, Stern PM (1987) Modification of a standardized system of wine aroma terminology. Am J Enol Vitic 38(2):143–146

Rainey BA (1986) Importance of reference standards in training panelists. J Sens Stud 1:149–154

Szczesniak AS, Brandt MA, Friedman HH (1963) Development of standard rating scales for mechanical parameters of texture and correla-tion between the objective and the sensory methods of texture evaluation. J Food Sci 28:397–403

11.3 For Instructors and Assistants

11.3.1 Notes and Keys to Successful Execution

1. Preparation of standards in these exercises requires a large time commitment. Make sure lab assistants have allotted sufficient time for obtaining materials and preparing samples. Note the time range for dairy product standards (24–48 h, not the same for all defects).
2. You may not wish to do all options in this lab. For example, if your school frowns on wine in class, the dairy products option offers an alternative. But remember the wines are only smelled, not consumed.
3. You may choose different references from those in the wine exercises by referring to the recipes in the original wine aroma wheel papers. Similarly, other dairy defects can be found in the book by Bodyfelt et al. or the updated edition edited by Clark et al. (2009), along with instructions as to how to prepare them. Feel free to contact the author if you have difficulty finding them.
4. Not all the texture references may be readily available. If you cannot locate them (rock candy may be hard to find), you can omit that scale point, but try not to skip two points that are adjacent.
5. Fortuitously, the 15 scale for sweetness parallels the percent wt/vol concentrations (e.g., 5 on the scale is 5 % sucrose per 100 ml total solution), until you get to the higher levels (above 10) where the sweetness function begins to saturate.
6. You should expect that the ratings for the fruit beverage samples may not match the sucrose references in terms of concentration. This is due to suppression of the sweetness by acid/sourness or other components. Thus, 5 % sucrose in Kool-Aid may be, on the

average, scoring less than a 5 on the sweetness scale. This is a discussion point.

7. Reinforce: These scales have nothing to do with hedonics!

8. You can discuss quality judging systems like the ADSA Collegiate Dairy Judging contest if students are interested. Stress that different defects are more or less damaging and have different points subtracted (also based on the intensity of the defect). Students in the contest are expected to memorize the point-subtraction (penalty) system for all defects.

9. As in previous labs with sucrose as percent weight to volume, you must be sure to arrive at the correct final volume. For example, for 10 % sucrose, you would take 10 g, add to about 50–75 ml H_2O, spin and/or stir until dissolved, then top off to reach 100 ml (stir again to mix). DO NOT simply add 10 g sucrose to 100 ml H_2O. The water will expand due to the partial volume taken up by the sugar, and you will end up with the incorrect final volume.

10. Most students will not be familiar with the mouthfeel of whole milk and may complain. This can be a discussion point, as trained panelists are not allowed to whine.

11.3.2 Equipment

For option 3, wine glasses with covers such as watch glasses. Ten per every five students, five for training and five for testing (40 glasses per 20 students).

For option 4, dairy defects, a stir plate, and 1 L flasks or similar glass containers are needed. Stove or hot/stir plate and thermometer for pasteurization if lipolized defect is made.

11.3.3 Supplies

Cups or plates for product presentation
For option 4, dairy defects, 6–8 oz. plastic cups with lids
Method for labeling (paper grocery store label gun or similar, or odorless markers)

Napkins, rinse cups, spit cups, unsalted crackers, and rinse water
Items needed for spill control and trash collection
Samples required (based on 20 students)

Option 1: Sweetness intensity references
5 lbs commercial cane sugar (2 kg)
Two–three packets unsweetened powdered drink mix

Make up the following solutions in 0.5 L batches per 20 students:
Scale point 2: 2 % sucrose in water (2 g per 100 ml final solution = 10 g per 500 ml)
Scale point 5: 5 % sucrose in water (5 g per 100 ml = 25 g per 500 ml)
Scale point 10: 10 % sucrose in water (10 g per 100 = 50 g per 500 ml)
Scale point 15: 16 % sucrose in water (16 g per 100 = 80 g per 500 ml)

Test samples: Make up 0.5 L batches of powdered, unsweetened drink mix (must not be "sugar free," which may contain an artificial sweetener, but totally unsweetened), as 5, 8, and 12 % sucrose wt/vol.

Option 2: Hardness texture references
Reference standards: Refer to Table 11.1 for the texture references. Note the sample sizes and multiply to obtain amounts for your class size. Kosher frankfurters were recommended for the original texture hardness scale. Rock candy was the high-end reference, but it may be difficult to find. If so, a Lifesaver-type candy may be substituted. The original samples were listed at ½ inch, which has been converted to 1.2 cm as an approximation. Sizes do not have to be exact.

Note that you will need one egg for each student, with the tip (white part) cut off after boiling (1 min to harden) and the rest of the egg discarded.

Test items: Chocolate chip cookie (1/4 per student, broken in four approx. equal size pieces), plain white bread (cut into 2″ squares), and celery stalk (1″ section per student) or similar vegetable

Shopping list (based on 20 students): 8–12 oz. cream cheese, 8–12 oz. processed cheese, 2 dozen eggs, one 6–8 oz jar stuffed olives, one 8 oz jar or

Table 11.1 Reference standards for texture hardness scale

Scale point	Item	Source/brand	Size	Serving temp. (recommended)
1	Cream cheese	Kraft/Philadelphia	1.2 cm^3	44–55 F
2	Egg white	Hard cooked	1.2 cm tip	Room
3	Hot dog	Large, skinless, uncooked	1.2 cm	50–65 F
4	Processed cheese	Kraft American or similar	1.2 cm^3	50–65 F
5	Olives	Large, pit-less, stuffed	1 olive	50–65 F
6	Peanuts	Shelled cocktail style, e.g., Planters	1 peanut	Room
7	Carrot	Uncooked, fresh	1.2 cm^3	Room
8	Peanut brittle	Candy part, no nut	1.2 cm^2	Room
9	Hard candy	Rock candy, or Lifesaver	~0.5 cm or 1 Lifesaver	Room

Table 11.2 Examples of reference standards for wine aroma terms

Term	Recipe	Alternative
Diacetyl (buttery)	1 drop butter flavored extract per 100 ml wine	
Oak	2–3 ml oak flavor per 25-ml wine	Fresh oak shavings, allow to steep overnight
Vanilla	1–2 drops vanilla extract per 25 ml wine	
Raisin	5–8 crushed raisins/25 ml wine	
Labrusca/methyl anthranilate	5 ml Welch's Concord grape juice	
Black currant	5 ml Ribena black currant juice	10-ml Cassis
Bell pepper	1 cm × 1 cm cut fresh green pepper, allow to soak 30 min before class and remove	

tin of cocktail style peanuts, 2 lbs peanut brittle (nuts will be removed/avoided), 3–4 fresh carrots, 0.5 lb rock candy or 2–3 rolls of Lifesaver-style hard candy, one pack (8 counts) all beef frankfurters (Kosher if available), 1 package chocolate chip cookies, 1 loaf plain white bread, and 1 bunch of celery

Option 3: Wine aroma references
Approx. 4 L white base wine. A plain smelling blended box wine with little varietal character is recommended or a nondescript varietal such as domestic Pinot Grigio. Refer to Table 11.2 for examples of defects.

Shopping list: One fresh green (bell) pepper, one small bottle vanilla extract, one small bottle of artificial butter flavor, one small can or bottle Ribena black currant juice, 1 package fresh raisins, two-ounce sample of Oak Mor, oak extract,

or similar, or one-foot section of 2-in. red oak for shavings. If items are difficult to obtain, you may omit one or two of these. Additional suggestions and recipes can be found in the second wine aroma wheel paper (Noble et al. 1987).

Option 4: Dairy product (milk) defects. Table 11.3 gives examples of dairy defects in fluid milk. You may choose four or five of them and also include a non-defective fresh milk sample.

2 L pasteurized whole milk for each defect.

If you decide to make the lipolized defect, you will have to obtain raw milk.

$CuSO_4 \cdot 5H_2O$ is needed (1 % stock solution) for the metallic catalyzed oxidation defect.

Ethyl hexanoate is needed for the fruity/fermented defect.

Fresh cultured buttermilk for the acid defect (half pint is sufficient).

Table 11.3 Examples of dairy defects and recipes for preparation

Defect	Recipe	Notes
Metallic oxidized	1.8 ml 1 % $CuSO_4$ solution to 600 ml whole milk	1 % $CuSO_4$ stock solution is prepared and stored
Light oxidized	Expose 600 ml whole milk to bright direct sunlight for 12–15 min	Prepare 24 h ahead, (no more—turns to met. oxid. after 2 days)
Rancid (lipolysis)	Mix 100 ml raw milk with 100 ml pasteurized milk and agitate in a blender for 2 min. Add 400 ml to obtain 600 ml total	Pasteurize after defect can be smelled by heating to 70 °C for 10 min. Prepare 24–36 h ahead
Fruity/fermented	1.25 ml of 1 % ethyl hexanoate (food grade) per 600 ml milk	
Cooked	Heat 600 ml milk to 80 °C for 1 min and cool	
Acid	Add 25 ml fresh cultured buttermilk to 575-ml whole milk	Prepare 24–48 h ahead. Taste to insure defect is perceivable

11.3.4 Preparation

General: Samples may be plated or placed on cups on trays ahead of time.

Reference standards should be clearly labeled with the intensity number, the name of the wine aroma or the name of the dairy defect.

All test samples should be labeled with random three-digit codes. Paper labels on cups are recommended. If a marker is used, it should be odorless.

Option 1: Be sure to make up sucrose standards and test beverages as % wt/vol. This means you must end up with the correct final volume. DO NOT simply add 10 g sucrose to 100 ml water, because the water will expand due to the volume taken up by the sugar molecules. You must start with a smaller amount of water, dissolve the sugar, and then "top off" to reach the final desired volume (and stir of course).

Reference standards and test samples may be served as 20 ml samples in 1–4 oz cups at room temperature.

Option 2: Hardness, texture scale
Serve reference standards in small cups, with the scale point label. Students should be instructed to expectorate the hot dog sample as it is uncooked. Serve the test items in three-digit coded cups or on plates divided into three sections and labeled.

Note the recommended serving temperatures. If it is not possible to maintain these temperatures, the samples may be served at room temperature, as long as that does not compromise the safety/integrity of the item.

Option 3: Wine aroma. For the wine aroma standards, you can use one set of training glasses and one set of test glasses for every five or so students. They can work in groups and pass the references around. They should feel free to discuss the references, but not the test samples (work independently). Several recommended references are given in Table 11.2. You may choose five or six of them if some items are difficult to obtain. Be sure to smell all prepared references and before class to make sure they are clearly perceivable. More than one person must smell the samples because of individual differences in olfactory ability (three smellers are recommended). If the sample seems very weak, adjust the recipe accordingly.

Option 4: Dairy defects. Pour samples as 50-ml samples in 100-ml plastic cups (4 oz or larger). Cups must have lids. If you make the lipolized defect, you should pasteurize the milk after the defect has developed and before serving. Heat to 70 °C for at least 10 min, then cool and store.

Samples should be prepared 24–48 h ahead of class to insure full development of the classic defect flavors. Prepare in glass containers. See pages 474–477 of the Bodyfelt book or pages 551–560 of the updated edition (Clark et al. 2009) for further information and details.

Acceptance and Preference Testing

12

12.1 Instructions for Students

12.1.1 Objectives

To familiarize students with the common methods for acceptance and preference testing.
To introduce options for dealing with no preference responses.
To reinforce simple *t*-tests on scaled data and binomial tests on proportions.

12.1.2 Background

The two most common methods for testing the consumer appeal of a food or consumer product on a blind basis are the methods using scaled acceptability and methods using preference choice. Acceptability refers to liking or disliking responses collected on a graded scale. The industry standard for measuring acceptability is the nine-point hedonic scale. The scale consists of four phrases for liking and four for disliking, using the modifiers slightly, moderately, very much, and extremely in front of the words like or dislike. A neutral point option is offered that is worded "neither like nor dislike." The data are usually scored and recorded on a one-through-nine basis, with nine being the highest degree of liking. This scale has a long track record, and in spite of criticisms and alternative scales that have been proposed over the years, remains in common use in the food industry.

An alternative procedure for assessing consumer appeal is based on choice, usually from a pair of products. The data consist of frequency counts of those people preferring each product. If a choice is forced (you must pick one), then the data can be analyzed by a simple binomial-based test against an expected proportion of 0.50 (50 % preference, or an equal split). This method is preferred among many research and development scientists as well as marketing managers because it is believed to be sensitive to differences and is easy to interpret. However, it has some deficiencies, namely, (1) one product may be preferred, but both may be disliked, i.e., poor offerings, and (2) a 50/50 split is ambiguous. It may mean there is really no preference, or there may be stable segments that truly like each version of the product offered in the test. Nonetheless, the simple paired preference test forms the basis for advertising claims of product superiority (e.g., "Taste the best hot dog in America!").

A further complication arises if the consumers are allowed to choose a "no preference" option. This third choice invalidates the use of the simple binomial statistical model, unless the data are massaged (or "fudged") to get back to two categories of frequency counts. Various options are discussed in Lawless and Heymann (2010), including (1) ignore the no preference votes (i.e., throw them away), (2) split them 50/50 among the two products, and (3) split them proportionally, based on those who did express a preference. That is, if the split is 60/40 among those expressing a preference, you allot the no preference votes in the same ratio. Other statistical treatments are available, including a multinomial analysis and Thurstonian modeling.

Both of these tests should only be done on samples of consumers who are frequent users of the product. Participants are almost always screened using a separate questionnaire to verify this requirement. For purposes of this lab, we will pretend that our class of students consists of all users of these products. As always, you may excuse yourself from tasting anything you would not normally eat for dietary, religious, or medical reasons.

Further information on acceptance and preference testing can be found in Lawless and Heymann, Chaps. 13 and 14. General considerations in consumer testing are outlined in Chap. 15, including questionnaire design.

12.1.3 Materials and Procedures

12.1.3.1 Materials
Four pairs of products will be presented that differ in their fat or sodium content. They will probably be presented on the same tray. If so, please make sure the three-digit codes match the ones on your ballots (you should always do this anyway).

12.1.3.2 Procedures
Half the class will do the acceptability testing first. The other half will do the paired preference tests first. In the acceptance test, taste each product and rate each product individually (without comparing or going back and retasting). Use the questionnaire with the nine-point hedonic scales. Choose one phrase that best represents your overall opinion of the product. When you have rated all four pairs of products, record your ratings on the master sheet.

In the preference test, taste the products in the order they are listed on the ballot (left to right). Circle the number of the test item you like the best (or dislike the least if you dislike both). If you have no opinion, like them both equally, or dislike them both equally, you may circle the "no preference" option.

When you have completed all four pairs, record your choices on the master sheet.

12.1.4 Data Analysis

1. Perform four paired t-tests on the scaled data. Report means and standard errors in a table or graph. Paired or dependent t-tests are illustrated in the first statistical appendix chapter in Lawless and Heymann (2010).
2. Perform binomial tests on the four sets of preference/choice data. The formula to use is the common test on proportions against a null/expected proportion of 0.5 as follows:

$$z = \frac{[P_w - 0.5] - \left(\dfrac{1}{2N}\right)}{0.5 / \sqrt{N}} \qquad (12.1)$$

where P_w is the proportion for the winning item, i.e., the one with a larger number of choices. N is of course the total number of judges in the count, and $1/2N$ is a continuity correction. z has a critical two-tailed value of 1.96 for $p < 0.05$.

Perform this simple binomial z-test on all four products using three different analyses of the no preference votes (1) ignoring the no preference responses (lowers N), (2) allotting them equally to the two choices, and (3) allotting them proportionally to the two choices based on the proportions of those expressing a preference. If you have a fraction when doing this, round up or down to the nearest whole person. If you have four pairs of products, this will require 12 calculations.

12.1.5 Reporting

This lab will use the standard format unless your instructor advises otherwise.

In reporting your results, do the following:
1. State which products were winners (=significantly higher ratings or significant preference split), if any, based upon your results.
2. Compare your conclusions from the preference test to your conclusions from the acceptability tests.
3. Make a table of the mean acceptability scores and standard errors for each product. Note whether the t-test showed a significant difference.

4. Make a table of your binomial z-test results. Include the z-value for each analysis method and each pair of products. Include a column of whether there was a significant preference (or not), and if so, list/show the preferred product.

5. If your class is large enough, you can make a table of preference choices by product usage frequency (include the no preference votes as a separate category). Check with your instructor. Put your table in an appendix to your report. Note any patterns that you see in your discussion. Are frequent users of the product different from nonusers or infrequent users in terms of their product choices?

6. In discussion, note any differences in the outcomes and decisions based on your different test methods and different analyses of the no preference votes.

12.2 For Further Reading

Angulo O, O'Mahony M (2005) The paired preference test and the no preference option: Was Odesky correct? Food Qual Prefer 16:425–434

Cardello AV, Schutz HG (2006) Sensory science: measuring consumer acceptance. In: Hui YH (ed) Handbook of food science, technology and engineering. Taylor and Francis, Boca Raton, FL. vol 2, Chapter 56

Lawless HT, Heymann H (2010) Sensory evaluation of foods, principles and practices, 2nd ed., Springer Science+Business, New York

McDermott BJ (1990) Identifying consumers and consumer test subjects. Food Technol 44(11):154–158

Peryam DR, Girardot NF (1952) Advanced taste test method. Food Eng 24(58–61):194

12.3 For Instructors and Assistants

12.3.1 Notes and Keys to Successful Execution

1. The choices of which products to include in this exercise are many. Variables of interest to food science students include factors like fat content and sodium content or products artificially sweetened vs. natural sugars. Another consideration is whether there is segmentation. For example, with orange juices, there are strong preferences for pulp levels (none, medium, and high are typical offerings these days). Another preference shows up in milk fat levels. Potato chips that are extra crunchy (so-called kettle chips) have a following. Segmentation may open a topic for discussion in your lab section. Make sure the products are discriminable.

2. Having a sufficient "N" for finding differences can be a problem. Obviously, a lab class of only 20 students does not have the same statistical power as a consumer test with 200 screened users. You can combine student data with that of previous years, if you keep a consistent product list from year to year. Alternatively, you can reproduce a given year's data. If this is done with random replacement, it resembles a bootstrap technique, a potential classroom discussion point.

3. If the class is large enough, you can require students to do a cross-tabulation of product preferences and usage frequency as a usage frequency question is asked at the bottom of the sample preference ballots. If the class is smaller than about 20 students, the numbers are probably too small to reach any conclusions, unless you have additional data from a previous year. You should stress that a frequency questionnaire is part of a routine screening procedure used to qualify consumers for participation.

12.3.2 Equipment

There is no special equipment for this exercise, other than normal kitchen and lab supplies.

12.3.3 Supplies and Products

See product suggestions below.
Cups or plates are needed for product presentation. Trays are useful.
Method for labeling (paper grocery store label gun or similar, or odorless markers).

Napkins, rinse cups, spit cups, unsalted crackers, and rinse water.

Items needed for spill control and trash collection.

Samples required/product suggestions — based on 20 students:

0.5 kg each low-fat and regular-fat cheeses of the same brand or style

1 L each of nonfat ("skim") and full-fat ("whole") milk

1 L each of low-fat and full-fat yogurts of the same brand or style

1 L each of low-sodium and regular vegetable juice (mixed vegetable or tomato)

Alternatives: regular and low-sodium snack chips such as potato chips or tortilla chips of the same brand or nuts (but beware of nut allergies!)

12.3.4 Preparation

Cut cheeses into 0.5 cm cubes, with 3–4 cubes per version per student placed in small labeled cups or on plates divided into labeled sections. Milks, yogurts, and vegetable juices should be given as 20-ml samples in 30-ml (1 oz) labeled cups or larger.

Samples may be plated or placed on cups on trays ahead of time and served at room temperature, unless to do so would create a safety issue or if it would make the product very different from its usual form (melted ice cream is obviously a poor idea). All samples should be labeled with random three-digit codes. Paper labels on cups are recommended. If a marker is used, it should be odorless.

12.4   **Appendix: Sample Ballots**

ACCEPTABILITY TEST

Date _____

TESTER ID NO._____

RATE EACH SAMPLE FROM 1 TO 9 USING THE SCALE BELOW.

SCALE:	9	LIKE EXTREMELY
	8	LIKE VERY MUCH
	7	LIKE MODERATELY
	6	LIKE SLIGHTLY
	5	NEITHER LIKE NOR DISLIKE
	4	DISLIKE SLIGHTLY
	3	DISLIKE MODERATELY
	2	DISLIKE VERY MUCH
	1	DISLIKE EXTREMELY

PRODUCT _____

SAMPLE # _____   RATING: _____

SAMPLE # _____   RATING: _____

PRODUCT _____

SAMPLE # _____   RATING: _____

SAMPLE # _____   RATING: _____

PRODUCT _____

SAMPLE # _____   RATING: _____

SAMPLE # _____   RATING: _____

PRODUCT _____

SAMPLE # _____   RATING: _____

SAMPLE # _____   RATING: _____

PREFERENCE TEST – SAMPLE BALLOT

Date _____

TESTER ID NO._____

PRODUCT _____

TASTE BOTH PRODUCTS AND THEN CIRCLE THE CODE NUMBER OF THE
PRODUCT IN THE PAIR THAT YOU PREFER. You (May / May not) choose the no
preference option.

PRODUCT_____   PRODUCT _____   no preference _____

Reason for preference: _____

How often do you consume products in this category? (check one phrase that best
describes your frequency of eating/drinking items similar to this product)

_____once a day or more
_____not every day but at least once a week
_____not every week but at least once a month
_____ less than once a month
_____ I never eat products of this type

PRODUCT _____

TASTE BOTH PRODUCTS AND THEN CIRCLE THE CODE NUMBER OF THE
PRODUCT IN THE PAIR THAT YOU PREFER. You (May / May not) choose the no
preference option.

PRODUCT_____   PRODUCT _____   no preference _____

Reason for preference: _____

How often do you consume products in this category? (check one phrase that best
describes your frequency of eating/drinking items similar to this product)

_____once a day or more
_____not every day but at least once a week
_____not every week but at least once a month
_____ less than once a month
_____ I never eat products of this type

PRODUCT _____

TASTE BOTH PRODUCTS AND THEN CIRCLE THE CODE NUMBER OF THE
PRODUCT IN THE PAIR THAT YOU PREFER. You   (May / May not) choose the no
preference option.

PRODUCT_____   PRODUCT _____   no preference _____

Reason for preference: _____

How often do you consume products in this category? (check one phrase that best
describes your frequency of eating/drinking items similar to this product)

_____once a day or more
_____not every day but at least once a week
_____not every week but at least once a month
_____ less than once a month
_____ I never eat products of this type

PRODUCT _____

TASTE BOTH PRODUCTS AND THEN CIRCLE THE CODE NUMBER OF THE
PRODUCT IN THE PAIR THAT YOU PREFER. You   (May / May not) choose the no
preference option.

PRODUCT_____   PRODUCT _____   no preference _____

Reason for preference: _____

How often do you consume products in this category? (check one phrase that best
describes your frequency of eating/drinking items similar to this product)

_____once a day or more
_____not every day but at least once a week
_____not every week but at least once a month
_____ less than once a month
_____ I never eat products of this type

Optimization by Ad Libitum Mixing and the Just-About-Right Scale

13

13.1 Instructions for Students

13.1.1 Objectives

To become familiar with the method of adjustment for optimization and the just-about-right scale.

To introduce the concept of contextual effects on hedonic decisions.

To learn the liabilities involved in optimizing products by benchtop adjustment alone.

13.1.2 Background

Many ingredients with a single or primary sensory attribute will possess an optimum level in a product, in which the consumer acceptability is maximized at some concentration or content level of that ingredient. A good example is sweetness as a function of sugar content. Foods can be too sweet, not sweet enough, or just about right. Food product developers need sensory information about that optimum level.

The method of adjustment for optimization is a procedure in which an ingredient is added or taken away from a food in order to match that food to a standard or to optimize it to the best possible taste. This is often accomplished with simple ingredients in homogenous products such as adding sugars or acids to a beverage. The beverage can be mixed (or adjusted) to taste, and the final level of sugars and/or acids can be measured

by refractometry, pH, and/or titratable acidity. The adjustment method is also known as "ad libitum mixing."

Results from the method of adjustment can be affected by contextual biases. The process of adding an ingredient and, in effect, "working up" toward a desired taste level may give the impression that the food has reached the best level before it actually has. Judgments are being made in comparison to the previously tasted levels that were weaker. Similarly, when diluting down from a more concentrated starting point, the formulator may be tempted to stop too soon because the current concentration seems about right in contrast to the initial one. Additional information about the method of adjustment, its uses, and pitfalls is found in Lawless and Heymann (2010), Chap. 9. Formulators need to be careful when adjusting at the benchtop to their own taste optimum—it is quite possible to "stop too soon," an effect called the error of anticipation.

Another method for optimization is just-about-right (JAR) scaling. Individual samples can be rated on a JAR scale as too weak, too strong, or just about right for an attribute such as sweetness or sourness. Data that are normally distributed around the center of the scale (i.e., the "just-about-right" point) are indicative of an optimized level of ingredient(s). It is important in this technique to examine the distribution of the raw data. For example, a consumer panel might contain one segment that prefers a soup with virtually no salty taste and another segment prefers a soup with a very salty taste. If this panel was presented with a soup with a moderate level of salt, the panel might produce a mean rating close to the "just-about-right" point. However, none of the raw data would actually fall at this point!

The just-about-right scaling technique is also prone to a centering bias. Given a series of concentrations, the middle concentration tends to be rated near the middle of the scale. Thus, the JAR scale may give a false impression of where the true just-about-right point is; it can wander a bit depending on the range of levels that are chosen for testing. Methods for estimating the true optimum point use two or more concentration ranges that are centered around different levels. The true just-about-right point is interpolated from the just-about-right estimates obtained from each series, using a graphical method.

This exercise includes optimization by the adjustment procedure and by a JAR scaling procedure. In the adjustment procedure, we will add a beverage containing a high concentration of sucrose to one that is more dilute until the dilute beverage seems to have about the right sweetness level (i.e., is optimized). We will also dilute a concentrated beverage and measure the ending sucrose levels in the optimized solutions using refractometry. Using JAR scaling, you will rate the sweetness levels of two different concentration series of a sucrose-sweetened beverage. The data obtained from the two series will be used to interpolate a true "just-about-right" point.

13.1.3 Materials and Procedures, Part 1: Optimization by Adjustment

13.1.3.1 Materials
Four cups containing a sucrose-sweetened powdered drink mix, labeled, and containing different sugar levels as follows:

200 mL of unsweetened powdered drink mix labeled with a three-digit code

200 mL of a highly sweetened powdered drink mix labeled with a three-digit code

"−−" 300 mL of unsweetened powdered drink mix

"++" 300 mL of highly sweetened powdered drink mix

13.1.3.2 Procedures
Begin with one of the two cups labeled with random three-digit codes (as indicated by your instructor or TA).

Taste the drink mix in this cup. Add a small amount of drink mix from the cup labeled "++" if you think the sample should be sweeter (to your own taste preference) or from the cup labeled "−−" if you think the sample should be less sweet (according to your own taste preference). Retaste the drink mix in the adjusted cup periodically until you believe it has the best tasting level of sweetness for you.

If you feel that you have added too much and overshot the optimal level of sweetness, you may add a small amount of drink mix from the "−−" cup or from the "++" cup until you have obtained the best sweetness level for you.

Once you have optimized the sweetness level in the first (random coded) cup, repeat the tasting and adjustment with the other cup labeled with a random code. When you are done with both mixing tasks, take your two optimized products to a TA so that the sugar content can be measured in each cup. The TA will measure the density of the solution in each cup using a refractometer, or students may be shown how to do this.

13.1.4 Materials and Procedures, Part 2: Just-About-Right Scaling

13.1.4.1 Materials
Powdered drink mix in two concentration series sweetened with sucrose. A dilute series consists of 2, 5, and 8 % sucrose (wt/vol), and a concentrated series consists of 8, 12, and 16 % sucrose.

13.1.4.2 Procedures
Obtain a ballot and six samples of the powdered drink mix from a TA. The samples are labeled only with random three-digit codes. Taste the samples in the order listed on the ballot and rate your perception of the sweetness level of each on the just-about-right scales. Take at least a 3-min break after tasting and rating the third sample. Decode the data on your ballot with the numbers 1–7 with the leftmost box ("not sweet enough") coded as 1 and the rightmost box ("too sweet") coded as 7.

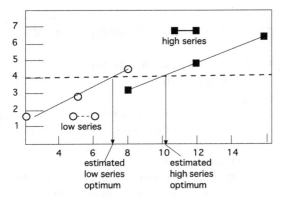

Fig. 13.1 Examples of interpolation from the high and low JAR series to find interpolated JAR estimates for further analysis

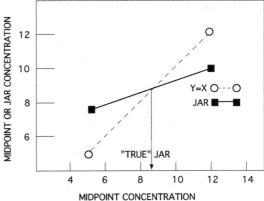

Fig. 13.2 Interpolation method for finding a true JAR point based upon the class average JAR estimates from the two series. The *open circles* show the midpoints of the series plotted against themselves ($Y=X$). The *black squares* show the two class average JAR estimates plotted against the midpoints. The interpolated point gives the true JAR point that would have been obtained if a single (hypothetical) series had been centered on the JAR point by a lucky guess. The centering bias effect is removed from the hypothetical centered series

Interpolate your individual just-about-right estimates for both concentration series using the blank graphs given out in class. An example of how to interpolate these points from your data is shown in Fig. 13.1. Give your ballot and your plots of individual just-about-right estimates to a TA for tabulation.

13.1.5 Data Analysis

Remember that you have two different sections of data from this lab:

Adjustment method. The optimization data where you mixed the solutions and they were read from the refractometers. One series of data is from the ascending optimization and the other from the descending (in sugar content).

JAR ratings. In class, you were to find your individual JAR by interpolating the concentration at JAR = 4, once for the high series and once for the low series. Find a class average for each point in each series, the lower series (2, 5, and 8 %) and a higher series (8, 12, and 16 %).

1. For optimization by adjustment, compare the ascending and descending optimization data by performing a paired t-test.
2. For optimization by JAR ratings, perform a t-test on the JAR data from the spreadsheet.
3. In the next plot, use a method of interpolation to find a true JAR estimate. It does this by finding a point at which the series would

have been centered on the true JAR, if we had known that fact ahead of time. Because the series would have been centered on the JAR point, the centering bias would be irrelevant. We are using the graphing method of Johnson and Vickers to "cancel out" the bias effect.

Plot as follows: Put the midpoint concentration on the x-axis. Draw two lines:

(a) First, plot the midpoints against themselves (5 vs. 5 % and 12 vs. 12 %). Connect the points. This should give you a line where $y=x$.
(b) Plot your JAR estimate from the low series (on the y-axis) against its midpoint (on the x-axis) and the JAR estimate from the high series against its midpoint.
(c) Drop a line down to the concentration axis from where the two lines intersect. This is your "true JAR" estimate, where the JAR point and the midpoint would have been in a balanced series centered on the true JAR point.

An example of this plot is seen in Fig. 13.2.

13.1.6 Reporting

Use the standard lab report format but combine your results and discussion (unless instructed otherwise). Be sure to answer the questions. Show your graphs and work.

13.1.6.1 Results and Discussion

1. Show your *t*-test results from the ascending and descending adjustment methods.

    Answer the following questions: Were the two directions different? How? Why do you think this happened?
2. Show the means and *t*-test results from the two JAR series.

    Answer the following questions: Were the two means from the dilute and concentrated series different? How? Why do you think this happened?
3. Report your interpolated JAR point. Answer the following question: How did the true JAR from the JAR scale ratings compare to the two values from the mixing (optimization) procedure?

    Append any calculations you made and remember to include your graphs.

13.2   For Further Reading

Johnson J, Vickers Z (1987) Avoiding the centering bias or range effect when determining an optimum level of sweetness in lemonade. J Sens Stud 2:283–291

Lawless HT, Heymann H (2010) Sensory evaluation of foods, principles and practices, 2nd ed., Springer Science+Business, New York

Mattes RD, Lawless HT (1985) An adjustment error in optimization of taste intensity. Appetite 6:103–114

Rothman L, Parker MJ (2009) Just-About-Right scales: design, usage, benefits, and risks. ASTM Manual MNL63, ASTM International, Conshohocken, PA

13.3   For Instructors and Assistants

13.3.1 Notes and Keys to Successful Execution

1. Any three-digit random codes may be substituted. If needed, the codes can be changed from year to year to discourage copying of previous lab reports.
2. The volumes given below are minimum amounts. Do not use any less. For improved accuracy, the adjusted cups can be increased to 200 ml and the supply cups from 300 to 400 ml. Remember the size of a typical male sip is 25 ml and the female sip about 15 ml.
3. It is important to start with *unsweetened* Kool-Aid or some other powdered drink mix. Beware of products labeled simply "sugar free" as they may already be sweetened with aspartame or some other intensive sweetener. Using such a product is a fatal mistake.
4. Sucrose percents are weight per volume, e.g., 20 % sucrose is 20 g of sucrose in 100-ml *final volume*. Do not add 20 g to 100 ml, but start with 20 g and mix slowly adding water to get 100 ml at the end. The water will expand as sugar is added due to the partial molar volume of the sucrose.
5. Larger batches are recommended based on the size of the class.

13.3.2 Equipment

Hand-held refractometers (2 or more) and trays.

13.3.3 Supplies

Dropping pipettes, deionized water, unsweetened powdered drink mix, sucrose (commercial grade), napkins, paper towels, cups, rinse water, and spill control items. Deionized water for cleaning refractometers between readings

13.3.4 Procedures and Sample Preparation, Part 1: Adjustment (Mixing)

For the optimization by adjustment, each student will receive four samples, a three-digit-coded cup with at least 100 ml of unsweetened powdered drink beverage, with no added sucrose, and a second three-digit-coded sample with 20 % wt/vol sucrose. A larger cup should be marked "--" and contain at least 300 ml of unsweetened powdered drink beverage (with no added sucrose) and a second cup with at least 300 ml labeled "++" containing the beverage with 20 % wt/vol sucrose.

Samples from the adjustment method should be served in ~120-ml cups (4 oz) for the items to be adjusted and large (500 ml or larger) cups for the reservoirs from which solution will be added.

As students are finishing their optimizations, they should label their cups with their initials or some unique identifier. Deliver to the TAs for refractometry or, if possible, you can instruct the students themselves to take the readings. It helps to have extra refractometers if students are taking their own readings. Train the TAs if necessary. Rinsing the lens between samples is key; bring deionized water for this purpose in a wash bottle. Extra dropping pipettes are useful for applying the sample droplets to the lens. Disposable pipettes are useful. If reused, these should also be rinsed between samples.

Data can be tabulated either as brix or as sucrose concentration if a calibration curve has been made. These are usually so close that there is little or no added benefit from a calibration curve, so it is optional. Tabulate results from the "diluting" series and the "concentrating" series. Students will be asked to do a paired t-test on the class raw data.

13.3.5 Procedures and Sample Preparation, Part 2: Just-About-Right Scaling

Each student will receive six samples of a powdered drink mix with three-digit codes in two series. The first is dilute series that consists of the beverage containing 2, 5, and 8 % sucrose (wt/vol) and the second is concentrated series that contains 8, 12, and 16 % sucrose.

13.3.6 Notes on Data Analysis

This is a moderately difficult exercise for some students, so you can anticipate questions and need to review the process. Provide blank plots for students to do their graphs and interpolations, similar to Figs. 13.1 and 13.2.

Not all students will be able to interpolate their own JAR points. Individual data may not form a straight line. Fitting "by eye" is acceptable during the lab period. Allow for students who may not be able to fit their data or find a sensible interpolation point. Some people do not like sweetness at all; others find the exercise hard to do.

Provide a key to determine which sweetness level was in which cup. Have students decode the scale so that "not sweet enough" = 1, "just-about-right" = 4, and "too sweet" = 7. They should then graph the results from each series using the provided (blank) plots or graph paper. Have them fit a line to the data in both of the plots. Fitting "by eye" is acceptable for this exercise. Next, they must interpolate from each line the sucrose concentration at which the "just-about-right" scale point lies (see Fig. 8.1 for an example). Do this for each series. Tabulate students' results.

13.4   Appendix: Sample JAR Ballots and Data Sheets

<u>Just-About-Right Ballot</u>

Taste each of the samples below, in the order specified, and rate each for its level of <u>sweetness.</u>

Sample 901

☐   ☐   ☐   ☐   ☐   ☐   ☐

Not sweet   Just-About-Right   Too
  enough   sweet

Sample 482

☐   ☐   ☐   ☐   ☐   ☐   ☐

Not sweet   Just-About-Right   Too
  enough   sweet

Sample 733

☐   ☐   ☐   ☐   ☐   ☐   ☐

Not sweet   Just-About-Right   Too
  enough   sweet

STOP... TAKE A THREE-MINUTE BREAK BEFORE CONTINUING WITH THE NEXT SERIES

Sample 629

☐   ☐   ☐   ☐   ☐   ☐   ☐

Not sweet   Just-About-Right   Too
  enough   sweet

Sample 494

☐   ☐   ☐   ☐   ☐   ☐   ☐

Not sweet   Just-About-Right   Too
  enough   sweet

Sample 135

☐   ☐   ☐   ☐   ☐   ☐   ☐

Not sweet   Just-About-Right   Too
  enough   sweet

Just-About-Right Ballot

Taste each of the samples below, in the order specified, and rate each for its level of <u>sweetness.</u>

Sample 733

☐   ☐   ☐   ☐   ☐   ☐   ☐

Not sweet   Just-About-Right   Too
  enough   sweet

Sample 901

☐   ☐   ☐   ☐   ☐   ☐   ☐

Not sweet   Just-About-Right   Too
  enough   sweet

Sample 482

☐   ☐   ☐   ☐   ☐   ☐   ☐

Not sweet   Just-About-Right   Too
  enough   sweet

STOP... TAKE A THREE-MINUTE BREAK BEFORE CONTINUING WITH THE NEXT SERIES

Sample 135

☐   ☐   ☐   ☐   ☐   ☐   ☐

Not sweet   Just-About-Right   Too
  enough   sweet

Sample 629

☐   ☐   ☐   ☐   ☐   ☐   ☐

Not sweet   Just-About-Right   Too
  enough   sweet

Sample 494

☐   ☐   ☐   ☐   ☐   ☐   ☐

Not sweet   Just-About-Right   Too
  enough   sweet

Part III

Brief Exercises and Group Projects

Group Exercise in Descriptive Analysis

14

14.1 Student Instructions

14.1.1 Objectives

To become familiar with ballot generation and descriptive analysis data collection.

To learn how to analyze data from a descriptive analysis panel.

To practice communication skills by producing a report and class presentation.

14.1.2 Background

In descriptive analysis methods, panelists, working individually, quantitatively specify the perceived intensities of a group of attributes that is specific to a particular product or class of products. The sensory attributes are chosen and refined by the panel in the ballot generation and the training steps when setting up a panel. So a critical process is the choice of terms to be included. This is often accomplished by a process of term collection in which all possible words that might be used to describe the product or product category are generated from a group of panelists or trainees. A panel leader will collect the words and organize the list by general categories such as appearance, aroma, flavor, texture/mouthfeel, and residuals. Further information on descriptive analysis techniques is found in Chap. 10 in Lawless and Heymann (2010).

Once the potential term list has been collected, the list is reduced and refined by eliminating those redundant or overlapping terms, those terms that are vague, and those terms that have an affective (i.e., like/dislike) meaning. Terms that have complex meanings or are complex combinations (e.g., creamy) are broken down into simpler components if possible. The term list continues to be narrowed, perhaps over several sessions, until the panel agrees on the meaning of each term and that the list of terms adequately describes the product of interest. During this process, the panel must choose anchor terms for the high and low ends of the scales on the ballot (e.g., "none" or "not at all_____" to "very_____"). Reference standards are often used to illustrate the meanings of the terms and sometimes to calibrate the panelists as to intensity levels. Panelists may participate by bringing potential reference standards to the group for discussion, or these may be suggested by the panel leader or by referring to the literature on sensory terminology and lexicons.

After the draft ballot has been constructed, the panel will be given practice sessions with various samples of the product to make sure that each member of the panel is using the attributes and scales in the same fashion. The inclusion of practice trials can eliminate errors such as misunderstanding term definitions, misunderstanding anchoring terms, or mentally reversing the scale. The ballot and list of terms may be further refined during this process, following group discussions and examination of statistical results. Standard deviations can be a helpful clue to the amount of group agreement.

Ratings usually are made on 15-point category or line scales. The scales represent intensity changes and never anything affective or hedonic. Data are analyzed using analysis of variance (ANOVA). Means, standard deviations, and standard errors are calculated for each attribute and product. Planned comparisons are then performed to determine where the means differ if the overall product F-ratio from the ANOVA is significant. Common tests on means include Duncan's multiple range statistic, Tukey's HSD, and Least Significant Differences (LSD) test. These are all modified versions of *t*-tests, enhanced to protect the chance of type I error when many tests are being made.

Once descriptive statistics have been calculated, the data are usually plotted so that the relationships between means and attributes may be visually compared. A frequently used graph for descriptive analysis is the "spider" plot (also known as a radar plot or radial-axis graph) that shows the means of several products on axes that radiate from a central point in the graph. Each axis (or radius or spoke) on such plots represents a single attribute in the descriptive analysis. If five to eight attributes are included on each plot, the mean ratings given a single product will be represented by a simple polygon in the graph. Products may then be visually compared by examining the individual shapes of two or more polygons on a particular graph.

14.1.3 Procedure

14.1.3.1 Specific Tasks and Workflow

1. Choose members—recruit friends if necessary to have $N \geq 8$ people.
2. Decide on product category—suggestion: a simple system with few attributes and some variation in the products that are out there.
3. Examine prior literature (Journal of Food Science, Journal of Sensory Studies, etc.) for terms. Many lexicons have already been developed for various products and these can be useful starting points.
4. Obtain samples of a few representative products—three to five.
5. Hold group discussions after tasting each product to generate terms for ballot. Review and refine to eliminate redundant, vague, affective, and complex terms.
6. Prepare actual ballots, complete with scales, attribute words, and intensity anchor words. Review finished ballot with TA or instructor for suggestions. Revise if necessary.
7. Conduct evaluation of 3–4 blind-coded samples from your category, using your ballot. These samples should include at least one unfamiliar product, e.g., brands not seen in the original ballot development, but from the same category.
8. Analyze statistically and prepare a report. ANOVA and suitable paired comparisons such as Tukey tests should be used.
9. Submit one written report from the entire group. The general format of objectives, background, methods, results, and discussion should be followed unless instructed otherwise. Cite any literature you used.
10. A class presentation of your results may be required. Check with your instructor or course website. Presentations should normally be targeted for a 15-min talk, including background, objectives, terminology, methods, results, and conclusions/discussion.

14.1.3.2 Additional Instructions

Term generation: The first product will be tasted and students will write on a blank sheet of paper all sensory properties that they individually perceive. Categories of sensations, such as appearance, aroma, flavor, mouthfeel/texture, and residual sensations, should be used to organize the list of terms and the final ballot.

The panel leader will then collect all of the terms from the panel in a group discussion. Group discussion will be used to eliminate redundant, vague, complex, or affective terms. Taste the second product and repeat the above exercise. The term list may be expanded upon or narrowed further with the information gathered from a second product. In addition, by sampling a range of products, the panel can begin to get an idea of what anchor terms should be used for each scale and whether or not physical reference materials are necessary to clarify the meaning of specific terms. The above exercise may be repeated using a third or fourth product to further refine the term list on the ballot.

NOTE: You must use the same people for the terminology development and for the evaluations. You should have a minimum of two sessions, one for term development and one for the formal evaluation.

Descriptive test: After your ballot is reviewed and approved by the instructor or TAs, you may begin the formal descriptive analysis. Choose three or four products from the same category you designed your ballot for. At least one of them should be something new that the panel has not seen in the training phase.

Collect your data and carry out the appropriate statistical analyses. This will usually involve ANOVA and planned comparisons among means such as the LSD test. Be sure to make a graph or other display of the means and some measure of dispersion such as the standard errors. The results should focus on the statistically significant differences. Do not waste a lot of time on attributes that are not significantly different.

Prepare a report and a class presentation if required. Check with your instructor or class website to see if a specific format is required. If a presentation is required, you should be able to talk about your choice of product and ballot/term development, report your results, and discuss any implications in a 20-min presentation.

You may wish to conduct the project using division of labor. One or more persons may be designated for individual tasks such as literature review, obtaining product, setting up samples, refining and printing the ballots, conducting statistical analysis, writing the report, and leading a class presentation.

14.1.4 Graphing Instructions

Here are specific instructions for graphing spider web plots (also known as radar plots) using Excel.

To get the clearest possible graph, it is a good idea to arrange the attributes so that highly correlated attributes are adjacent to one another. An easy way to do this is to look at results from a principal component analysis (PCA) if one is provided. By going clockwise or counterclockwise in terms of the attributes, correlated attributes will be adjacent, and in the spider web plot, not many of the lines will cross. This will make differences in the shape of the polygons easier to see.

The data below came from one of the wine laboratories at UC Davis. You may copy these data into your own Excel file or enter them by hand if not provided by your instructor or website.

Use Radar plots. The following uses Excel 2003 for Windows/PC. If you have a more recent version of Excel or are using a Mac, these options may have changed.

Enter your data as shown in the table or download it from your course website if available.

Highlight the columns wine to viscosity and the rows wine to arneis.

Now click on INSERT→CHART→Radar.

Choose the style of the subchart that is located in the most left-hand block.

Click on NEXT; make sure that the series is in ROWS.

Click on NEXT; label the chart.

Click on NEXT; place the chart in a NEW SHEET with a name.

Click FINISH.

This version of the graph is shown in Fig. 14.1.

You can further improve the format of the graph by clicking on the various lines, axes, and legends.

You can CLEAR the GRIDLINES, and FORMAT the DATA SERIES by changing the LINE STYLES and WEIGHTS. Then FORMAT the CATEGORY LABELS by making the FONT larger and by using BOLD. The legend box may be moved and made a little larger.

This version of the graph is shown in Fig. 14.2.

14.2   For Further Reading

Lawless HT, Heymann H (2010) Sensory evaluation of foods, principles and practices, 2nd ed., Springer Science+Business, New York

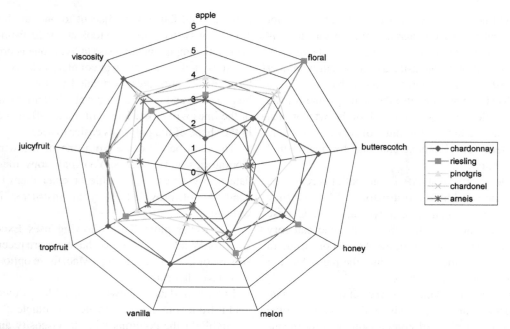

Fig. 14.1 Sample radar plot from the data in Table 14.1, made in Excel 2003

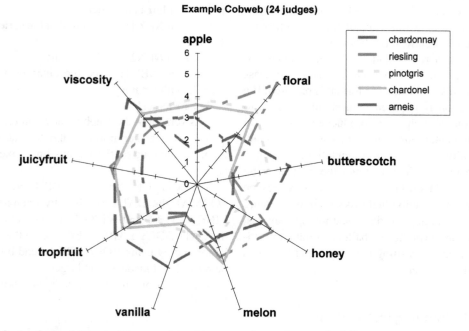

Fig. 14.2 Another sample plot of the same data with some options to improve legibility

Table 14.1 Sample data for constructing a radar plot

Wine	Apple	Floral	Butterscotch	Honey	Melon	Vanilla	Tropfruit	Juicy fruit	Viscosity
Chardonnay	1.4	2.9	4.5	3.5	2.6	4	4.4	3.9	5
Riesling	3.2	6	1.7	4.2	3.5	1.5	3.6	4.1	3.3
Pinot gris	3.9	4.4	3.5	2.3	2.7	2.2	3.1	3	4.2
Chardonel	3.6	4.2	1.6	2.7	3.8	1.9	4	3.9	4.1
Arneis	3	2.4	1.8	2	2.8	1.4	2.6	2.6	3.8
LSD	1.5	1.4	1.6	1.3	None	2.1	None	None	0.9

Lawless HT, Torres V, Figueroa E (1993) Sensory evaluation of hearts of palm. J Food Sci 58:134–137

Meilgaard M, Civille GV, Carr BT (2006) Sensory evaluation techniques, 4th edn. CRC Press, Boca Raton, FL

Stone H, Sidel J, Oliver S, Woolsey A, Singleton RC (1974) Sensory evaluation by quantitative descriptive analysis. Food Technol 28: 24–29, 32, 34

14.3 For Instructors and Assistants

14.3.1 Notes and Keys to Successful Execution

1. It is useful to hold a practice session with the entire class to act as a panel leader in terminology development. Choose a simple product such as apple or grape juice and conduct a term generation session with two or three samples. If resources permit, the mock ballot can be used in a subsequent week to illustrate the actual evaluation session.

2. Simple products should be chosen. Complex or difficult to evaluate products such as chocolate or cola beverages are generally poor choices. These can lead to a lot of frustration with a long ballot but few significant differences. Students may wish to evaluate differences in some variables of technical interest such as a fat-reduced version and sodium-reduced version of the product chosen. A variable of interest can provide additional motivation.

   NOTE: Whether to allow students to choose an alcoholic beverage for their product can present some challenges. Alcoholic beverages can be problematic unless the class is taught specifically in an enology setting. Such products require justification and should only be done by legal age panelists, under supervision, stressing expectoration and with correct tasting methods and a responsible professional attitude.

3. If a sensory software system such as Compusense or FIZZ is available, students may be trained to use it for entry of their ballots and actual evaluations. The system may also provide statistical analysis and the instructor should decide whether this is a valuable time saver or whether it is more useful to have students do their ANOVAs using some other statistical package or "by hand."

4. The project is well suited to having students present results to the class. Depending upon the size of the class, this can take an entire lab period or longer. It provides an opportunity for practice in presentation and verbal communication skills. Many students will already have some experience with PowerPoint or some other presentation software.

5. As in any group effort, some students will do more work than others. The assignment of a group grade (equal for all students) can sometimes raise complaints about unequal contributions, especially if a student slacks off and does little or nothing to contribute. Dealing with such individuals can be a valuable life lesson. One approach is to have students evaluate each other's contribution or give a partial grade (anonymously).

6. Class time may be allotted to the project or students may work outside of class. If the sensory test facility is required, students should of course schedule time slots. Be wary of students who try to conduct their sessions in

dorm areas or such. If work is done outside of the sensory lab or lab classroom, students should be required to submit pictures of their ballot development sessions and evaluation to insure that the correct procedures were performed with a professional attitude.

7. The recruitment of friends to bolster the panel size can present a challenge. One option is to have students participate in more than one group—one as workers and the second merely as panelists. Another option is to recruit from other students who are not taking the class but who are within the major as they may be interested in the sensory testing anyway. A common faulty shortcut is for students to use their group members for the term generation and then recruit others for the actual product test phase. This should be discouraged (use the same panelists in both phases).

8. Reports from previous years may be kept for illustration, with names removed, as examples of good/poor reports. Students find this helpful in determining your expectations as this kind of project is complex and probably new to them. A scientific journal format is suggested for the reports. A simple example can be found in the article listed in the References below.

9. Timing. As in any extended project, students should be warned about the time required. To avoid procrastination and crisis as the report deadline approaches, a progress report may be required for the term generation and ballot development phase to make sure it is not left unattended until the deadline.

14.3.2 Suggestions for Evaluating Class Presentations

Class presentations should be graded by at least three evaluators, who should be given an evaluation sheet with fixed criteria. Criteria should be made available to students ahead of time. It is valuable to have another faculty member, senior sensory lab personnel, or upper level graduate students come to the presentations, ask questions, and act as evaluators. A point system, such as one to four points, can be used for categories such as the following:

Introduction: Did they justify their choice of product? Did they discuss any previous literature?

Methods: Did they describe their method including the ballot development process and choice of terms? Any difficulties or problematic terms? Did they show the ballot or a table of terms and anchor words? Were there specific instructions as to how the product should be tasted, smelled, viewed, etc.?

Results: Did they focus on describing the product differences or just report statistics? Did they waste time on nonsignificant effects?

Graphics and visuals: Were the data appropriately illustrated with radial plots or bar graphs or similar? Were slides legible with suitable font size and well organized?

Discussion: Did they discuss the results beyond the statistics?

Q&A: Did they answer questions appropriately? Were they able to answer with "I don't know" or "I don't know but I'll find out and get back to you" or was there a futile attempt to cover what they didn't know?

Overall: Was there good organization? Did they start and finish on time? Did they bring examples of the products to show and/or pictures of the product in package?

15.1 Instructions for Students

15.1.1 Brief Exercise 1: Developing a Procedure for Meat Discrimination Tests

15.1.1.1 Objectives

To develop a test procedure for a difficult product.
To learn to specify the details of a sensory test.
To introduce difficulties in triangle testing.

15.1.1.2 Background

The two primary requirements for sensory evaluation are blind presentation and the principle of collecting independent judgments. Beyond these concerns, a sensory testing procedure has many details that must be attended to in performing a valid and reproducible sensory evaluation. Many of these details are found in Chap. 3 in Lawless and Heymann (2010). Various other guides have been published over the years from ASTM Committee E18 and the Institute of Food Technologists which has specified the details that must be included in published papers using sensory test methods. Many food and consumer product companies have books of standard operating procedures ("SOP" specifications) that describe in detail the procedures used in that laboratory or company.

The items to be specified can be broken down into four categories.

The first category concerns the people. Who will be in the test? What are their qualifications? How will they be found, recruited, screened, and trained or oriented if necessary? Are there any grounds for exclusion?

The second section concerns the samples. How will they be obtained or fabricated? What is their exact form or composition? How will they be prepared, handled, presented, and disposed of if unused? What are the details with regard to sample size or volume, serving temperature, and blind coding?

The third is the procedure itself. This should be detailed enough so that a person who is not highly educated in sensory test methods could follow it exactly in a remote location with little supervision. This should also specify what responses are required from the participants, what the ballot or questionnaire should look like, and whether the method is a standard procedure or some other common method referred to in the literature (e.g., triangle test). A critical concern is sample presentation order. Will the orders be randomized or counterbalanced? A standard design such as a Latin square or Williams design may be specified. Replication should be considered.

Special instructions may be given to panelists regarding methods of tasting and whether rinsing the mouth between samples or using provided palate cleansers is mandatory or not. In certain types of studies, it is also necessary to control the rate at which samples are tasted (e.g., in studies of hot or spicy foods where carryover effect is common, or certain tastes may linger for a period of time).

H.T. Lawless, *Laboratory Exercises for Sensory Evaluation*, Food Science Text Series 2,
DOI 10.1007/978-1-4614-5713-8_15, © Springer Science+Business Media New York 2013

The fourth major category concerns the data handling and statistical analysis. How will the responses be analyzed to test for statistical significance? What are the null and alternate hypotheses and probability values for significance cutoffs? What common actions are taken if the data show significance or what is recommended if they do not?

Other details may include equipment and supplies, sources for raw materials, the type of water used in the preparation, and other details if not specified in other sections. If a software system for sensory testing is used, the system and version number should be stated. Safety precautions should be spelled out (gloves, hair nets, lab coats, fire and spill control, etc.) unless they are another part of standard laboratory training. Finally, institutional review for the protection of human subjects and confidentiality needs to be specified, or the justification for exemption from IRB review should be stated.

15.1.1.3 Instructions

Form groups of three to four people. Discuss the following scenario and prepare a short lab report (to be turned in individually) from your group discussion. Your report should specify the test procedure in as much detail as you can give.

Scenario: A manufacturer of a pork shoulder, shelf-stable meat product wishes to introduce low-fat and low-sodium versions of their normally very fatty, very salty product. Consumers ordinarily prepare the product by cutting it into strips and frying it.

15.1.1.4 Report

This report is not in the standard format for an experiment, but concerns only the methods. Design triangle tests to determine whether or not differences exist between the low-fat and full-fat products and whether or not differences exist between the low-sodium and the standard salt products. A triangle test consists of three products, one pair of duplicates and one different sample, and the tester is asked to choose the sample most different from the other two. Each student should submit an individual report specifying the procedure under the following headings:

Participants, Samples, Procedure, Data handling and Analysis, Other details.

Work under the following assumptions:

1. No computerized data collection system will be used (paper ballots or questionnaires).
2. You will be giving detailed instructions to a technician at another site; it will be the technician that actually conducts the tests.
3. The technician was a chemistry major and a recent grad with no sensory testing experience.

List all of the experimental details that will be important to monitor and control. Also describe exactly how the product will be prepared and presented in the tests. Be as detailed and specific as possible. Assume that the procedure you write here will be entered into a lab procedure book as a standard operating procedure (SOP). These directions will be followed by others when performing triangle tests on this type of pork shoulder product. Consider whether special lighting might be needed to disguise visual differences.

Include the following in your report and any other details you think are important:

Sample coding and labeling

Presentation orders (how determined?)

Food sample size and serving temperatures

Cooking method

Holding and handling method (including any serving materials/utensils used)

Ballot/questionnaire format (including specific instructions to panelists; will retasting be allowed?)

Timing between samples and any other timing that could be important

Methods of data tabulation

Testing room conditions and lighting, air handling

Panelist screening methods: Who are the panelists? How will they be recruited? Anybody excluded?

Panelist motivation to participate (incentives?)

Panelist and technician safety considerations and kitchen and food handling issues

Chapter 3 in Lawless and Heymann (2010) is helpful in considering some of the procedural details and Chap. 4 discusses the basics of discrimination tests such as the triangle test.

15.1.2 Brief Exercise 2: Probability and the Null Hypothesis

15.1.2.1 Objectives

To explore the concepts of random chance, the null hypothesis, and probability values in choice behavior.

15.1.2.2 Background

Almost all statistical tests start from an assumption that only chance variation is operating in an experiment. This may seem like a silly idea, because we often design experiments with treatments or changes in a product that we know will have some effect (perhaps of a sensory nature). But this is the way experimental science works. So we start with an assumption that the treatment has no effect, and any data that we see that indicates some difference from what's expected by chance is just a random event. When we see some data that are so unlikely under this assumption (called a null hypothesis), we reject that idea and conclude that our treatment did have an effect, i.e., that our product changed in some systematic way.

This lab will explore the effects of random chance in a brief experiment on extrasensory perception (ESP). We will use a famous deck of cards invented by the psychologist J.B. Rhine called Rhine cards. They have five symbols on them, and your job is to guess which symbol your partner is looking at. If you do much better than chance at this, we might be tempted to conclude that you have ESP!

Because there are five symbols, we should expect your guesses to be correct about 1/5 of the time. Over the deck of 25 cards, we would expect you to get about 5 correct. However, due to random events, some people will do better than chance (and some perhaps worse). The distribution of random events with just two outcomes (correct or incorrect choice) is described by the binomial distribution. The binomial expansion that shows how the exact probabilities are calculated for various outcomes is shown below [extra credit option, (15.2)]. There is also a simple z-score formula for the normal distribution approximation to the binomial [see (15.1)].

If we see a level of performance that would be expected by chance only 5 % of the time or less, we would conclude, scientifically, that chance was probably not operating, that something (spooky?) was going on. But you might want to ask how many times such a result would be obtained if we tested everyone in class. How would you adjust for the fact that you are making many observations over many people?

15.1.2.3 Instructions

Form pairs of students. If you are by yourself, ask one of the TAs to be your partner. One student will act as the "sender" and the other as "receiver." The sender should shuffle the deck of Rhine cards and the receiver should hold the index card showing the five options (square, circle, cross, star, wavy lines). The sender should place the deck face down on the table and lift the first card so the receiver cannot see it. The sender can stare at the card if he or she wants, but should be careful that the receiver cannot see it reflected if sender is wearing glasses. After a 5 s wait, the receiver should guess which symbol is shown. The sender should record the answer as correct or incorrect. Record the total amount correct at the end of deck and give the TA your total. The sender and receiver should then switch positions and repeat the procedure. The class data will be posted or sent to you by e-mail.

15.1.2.4 Report

There is no specific format for this report. Answer the following questions:

1. What number correct out of 25 would be expected by chance? What number correct out of 25 would be expected to occur 5 % of the time or less? To find this, you can use the binomial approximation to the normal distribution z-score where

$$z = \frac{[(X/N) - p] - \frac{1}{2N}}{\sqrt{pq/N}} \quad (15.1)$$

where $z =$ the cutoff in the tail of the normal distribution to indicate a level seen only 5 % of the time $= 1.645, N = 25, p = 1/5, q = 1 - p = 4/5$

and the proportion observed $= X/25$ where X is the number correct that satisfies this equation. In other words, solve for the proportion observed $(X/25)$ that would give you a z-score of 1.645. Show your calculations.

2. Did anyone in class achieve this level? How many times would we expect performance better than chance at the 5 % level in a class this size?
3. Was there any evidence, in your opinion, for ESP from any class member?
4. Was there any evidence, in your opinion, for ESP from the class as a whole?
5. What is the null hypothesis for a test with a chance probability of 1/5?
6. What is your alternative hypothesis?

Extra credit exercise (optional): Using the binomial expansion, show the exact probability of the number correct you found above. Show your calculations. For the probability of an event x out of N possible occurrences, the probability is given by the following expression (remember to sum!):

$$p(X) = \frac{N!}{X!(N-X)!} p^x (1-p)^{N-X} \quad (15.2)$$

15.1.3 Brief Exercise 3: Consumer Questionnaires for a Military Field Ration

15.1.3.1 Objectives

To develop a questionnaire suitable for a consumer field test.

To introduce the topic of special foods for special circumstances.

15.1.3.2 Background

Food technology plays a major role in the development of foods for special purposes. Good examples are foods as nutraceuticals, foods for disaster relief, and foods for military field rations or for special circumstances such as space flight. This exercise will involve a pilot field test with a field ration developed for soldiers' use on patrol or when a field kitchen is not accessible for the preparation of normal meals. The ration is called

a Meal, Ready to Eat or MRE for short. The MRE has been a mainstay of the US military for about 25 years, during which time it has undergone many changes and technical advances. The ration is designed to deliver about 1,300 cal with a good portion of this energy derived from carbohydrates, for physical and mental performance. Typically, the MRE will contain an entrée, starch or vegetable, a snack item, dessert, and beverage. It also has condiments and a heating device.

In today's exercise, you will each be given an MRE to prepare and sample. A draft questionnaire will be filled out.

NOTE: The heating element may give off hydrogen gas. The product must be prepared outdoors. Do not smoke or put any kind of flame near the heating pouch after the water is added. Follow the directions carefully.

You will need: Water, approximately 500–1,000 ml for beverage prep and to activate the heating unit (NOTE: that water must be discarded). Some recent versions of the MRE contain a water packet used to activate the heating element. A "rock or something" to lean the heating pouch against

15.1.3.3 Instructions

1. Form pairs of students. There is one MRE for every two people. Prepare and sample the MRE. Read all the instructions and packaging carefully!
2. Develop a questionnaire, using the principles outlined in Lawless and Heymann (2010) Chap. 15. The questionnaire should be the kind used to compare two products in a monadic sequential central location test and it should address all of the following issues:
   (a) Acceptability of the entrée (main dish you heated)
   (b) Overall sensory acceptability of the meal
   (c) Overall satisfaction with the meal
   (d) Ease of use and preparation

15.1.3.4 Report

The standard lab format is not required for this lab. Turn in your finished questionnaire. It should be typed and include all instructions necessary for the test personnel and/or the consumer.

Extra credit (optional): Devise and describe a method for administering a questionnaire under harsh conditions (summer desert or winter mountain warfare exercise). Your test method should enable persons to complete the questionnaire under less than ideal, stressed circumstances (like trying to keep from freezing to death). You have an almost unlimited budget to develop a device, technique, or mechanism for this data collection. If you need to modify the questionnaire from class, describe what needs to be changed. Describe your invention to the best of your ability. One page is sufficient.

15.1.4 Brief Exercise 4: Shelf Life Estimation

15.1.4.1 Objectives
To become familiar with methods for shelf life measurement and stability testing.
To understand some methods for handling time-dependent data.

15.1.4.2 Background
Shelf life or stability testing is an important part of quality maintenance for many foods. It is an inherent part of packaging research because one of the primary functions of food packaging is to preserve the integrity of a food in its structural, chemical, microbiological, and sensory properties. A good review of shelf life testing can be found in the packaging text by Robertson (2006) with information on modeling and accelerated storage tests. A brief summary of shelf life measurement is found in Lawless and Heymann (2010) in Chap. 17. For many foods, the microbiological integrity of the food will determine its shelf life, and this can be estimated using standard laboratory practices; no sensory data may be required. The sensory aspects of a food are the determining factors for the shelf life of foods that do not tend to suffer from microbiological changes such as baked goods. Sensory tests on foods are almost always destructive tests, so sufficient samples must be stored and available, especially during the period in which the product is expected to deteriorate.

Shelf life testing may employ any of the three major kinds of sensory tests, discrimination, descriptive, or affective, depending upon the goals of the program and resources available. Thus, one can view shelf life tests as no special category of sensory testing, but simply a program of repeated testing using accepted methods.

Two main choices are used for criteria for product failure. These include a cutoff point and statistical modeling with equations such as a hazard function or survival analysis. Note that product failure is an all-or-none phenomenon, and decreases in sensory measures such as falling acceptability or increasing percents of consumer rejection are more continuous in nature. This opens the opportunity for other kinds of models, such as logistic regression of time against the proportion of consumers rejecting. In this exercise, we will look at proportion rejecting and plot it using probability paper and two time scales, arithmetic and logarithmic. An alternative cutoff point can be determined using a descriptive analysis panel and/or instrumental measurements of texture and flavor changes. However, descriptive panel data should be calibrated against consumer acceptance information in a separate study to determine how much of a change is critical.

Failure occurrences are not always normally distributed. Sometimes there are a few products in the sample set that last a very long time relative to the mean failure time. An example is light bulbs. Therefore, the failure distribution is often positively skewed or log normal. For this reason it is common to model shelf life as a function of log time, rather than simply time (usually in days for a food product). Because we have binomial data in this lab (failed or not failed), a logistic regression is an appropriate choice. Logistic distributions are similar to the normal distribution except that they are slightly "heavier in the tails."

15.1.4.3 Procedure
You will be shown various pictures of bananas and asked to judge each one on the following options: OK = I would probably eat this fruit. X = I would probably not eat this fruit.

15.1.4.4 Analysis

The instructor will tabulate the percent rejecting and provide the class data. There are two ways to get your estimate of shelf life:

1. Plot the proportion rejecting against time on probability paper and against log time on log probability paper. Show your graphs.
2. Fit a logistic regression equation against time and against log time. Once you have the equations, you can interpolate and find the 50 % failure time.

The logistic regression equation takes the form

$$\ln\left(\frac{p}{1-p}\right) = b_0 + b_1 X \qquad (15.3)$$

where p is the proportion rejecting and X is either time (in days) or log time.

Once you have your slope and intercept estimates (b_0 and b_1), you can solve for X at $p = 0.50$.

Find the 50 % point as an estimate of the shelf life in days.

15.1.4.5 Report

Answer the following questions:

1. Do you think the 50 % point is a good measure? Why or why not?
2. What kinds of other mathematical functions have been fit to survival analysis data?
3. Did your linear or log plot seem to show a better straight line?
4. What is the "bathtub" function and what does it describe?

Turn in your graphs, your 50 % estimates, and the answers to the three questions.

15.2   For Further Reading

Hough G (2010) Sensory shelf life estimation of food products. CRC Press, Boca Raton, FL

Hough G, Langohr K, Gomez G, Curia A (2008) Survival analysis applied to sensory shelf life of foods. J Food Sci 68:359–362

Lawless HT, Heymann H (2010) Sensory evaluation of foods, principles and practices, 2nd ed., Springer Science+Business, New York

Robertson GL (2006) Food packaging, principles and practice, 2nd edn. CRC Press, Boca Raton, FL

15.3   For Instructors and Assistants

15.3.1 Brief Exercise 1: Developing a Method for Meat Discrimination Tests

15.3.1.1 Notes and Keys to Successful Execution

1. The instructor or assistants should prepare some of the product in front of the class in order to illustrate, visually, how the product is cooked.
2. The exercise historically has used the Hormel product, Spam®, as the test item. It is currently available in "light" and low-sodium versions as well as the traditional formula. Other products may be substituted and should be considered if a large number of students find pork to be objectionable on religious or other grounds. However, students need not consume any of the products in this lab. Tasting may be optional.
3. Meat products are notoriously difficult to deal with due to the inherent animal-to-animal variability (eliminated here to some degree by using a comminuted meat product) and difficulties in consistent cooking, temperature, and holding times. Careful attention should be paid to the issue of whether the product can be cooked to a consistent internal temperature or whether time of cooking is a sufficient specification. Note that it is nearly impossible to put a meat thermometer into a thin slice of Spam, so that cooking to temperature is rarely a viable option without special thermodes or thermocouples.
4. It is assumed that students understand the basics of discrimination tests, the triangle procedure itself, and the statistics involved.
5. Additional materials. This lab can be entertaining if the instructor brings additional materials. The Hormel website has historical information on Spam. There is also a Spam cookbook, developed in the UK in response to wartime use of Spam and the associated post-war nostalgia over the dishes people thought up. Trivia is useful. Hawaii is the largest consumer (per capita by state) and Spam pizza is common. There are websites with Spam haiku, and the MIT website has over 19,000

poems. A comedy routine in the TV show, Monty Python, featured Spam.

6. Web resources
    (a) http://mit.edu/jync/www/spam/archive.html
    (b) http://en.wikipedia.org/wiki/Spam_Monty_Python
    (c) http://www.hormel.com

15.3.1.2 Equipment

Electric fry pans are a good choice for class demonstration. An alternative is individual propane burners and nonstick fry pans or skillets. Cutting boards and knives are required. Special slicing devices for uniform slicing may be illustrated if available.

15.3.1.3 Supplies

One can each of Spam, Spam light, and low-sodium Spam.

Vegetable oil, food handling gloves, hair nets, and paper or plastic plates for serving.

Water, spill control, napkins, spit cups, and plastic forks or similar.

15.3.1.4 Suggested Grading

Student work may be graded on the basis of the completeness and detail of their test specification as well as the reasonableness of their methodological choices. Although students may work together in groups during the lab period to develop their rough outline, they should submit individual reports unless the instructor is willing to assign group grades for single group reports. Because of the group effort during the lab period, there are special liabilities (opportunities) for shared writing and plagiarism in the submitted reports, so the expectations of individual work should be clearly spelled out. Some commonalities in the reports are to be expected.

15.3.2 Brief Exercise 2: Probability and the Null Hypothesis

15.3.2.1 Notes and Keys to Successful Execution

1. This lab is designed to show the students that there is variability around the chance perfor-

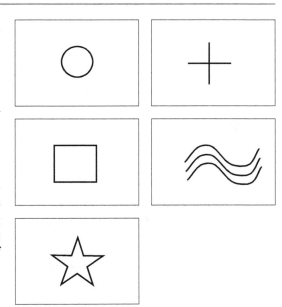

Fig. 15.1 An example of a card deck used by J.B. Rhine to study parapsychology, as originally designed by Karl Zener

mance level in a choice test. The use of an ESP test is only to provide some degree of student interest in what is otherwise a tedious exercise.

2. Samples of the cards are shown in Fig. 15.1. Common index cards work well.

3. Students should be aware that if there are 25 or more students in the class, chances are that one or more of them will obtain a performance level significantly above chance at the alpha = 0.05 level.

4. Note that the null hypothesis is that the population proportion correct is 1/5 and the alternative hypothesis is that the population proportion correct is greater than 1/5. So, as in common discrimination tests, the test is one-tailed, and thus, the z-score cutoff is 1.645, not 1.96. Confusion arises because the "population proportion" is sometimes specified by the letter "p," as is the chance probability level.

5. This is an opportunity to reinforce the idea that the null hypothesis is not that there is "no difference" or that "the result was due to chance" or any other such verbal statement. The null hypothesis is a mathematical equality, not a verbal statement. The alternative hypothesis is a mathematical inequality.

15.3.2.2 Equipment
None.

15.3.2.3 Supplies
Card decks, one for every two students. An example of a card deck as used by Zener and Rhine is shown in Fig. 15.1.

A tabulation sheet should show the number of correct guesses for each student.

15.3.3 Brief Exercise 3: Consumer Questionnaires for a Military Field Ration

15.3.3.1 Notes and Keys to Successful Execution

1. MREs may be obtained from outdoor sporting goods supply stores or survivalist mail order suppliers. These are almost always identical in components to the military forms but be careful to read the product claims. They may differ in condiments such as the hot sauce, and the external package is likely to be flimsy. The heating elements may or may not be included and it is important to order the meals WITH heating elements. MREs will typically be obtained in mixed cases with different main dishes, so students may be encouraged to compare the contents of different packets.
2. Due to the gases given off by some of the heating elements (e.g., hydrogen), the meals should be prepared outdoors. This is an opportunity to observe your students' performance and demeanor outside the classroom setting.
3. The optional exercise will be more meaningful if the outdoor preparation is done in bad weather (cold, rain, snow). This lab exercise was routinely done at Cornell University in late November, sometimes in snow. The lab was originally inspired by an actual field exercise in the US Marine mountain warfare training, conducted in the Sierra Nevada mountains in January 1979, in which food tech personnel were required to follow the Marines on snowshoes for several days and collect questionnaires in life-threatening subzero conditions. The optional exercise arose from the difficulty of that study. Class discussion may be held

indoors after the meals are prepared and sampled if necessary.

4. It is expected that many questionnaires will appear similar in content and format. Students should be made aware of the expectation for individual work unless this is done as a group exercise.
5. This lab can reinforce the importance in food technology of (1) packaging materials and (2) extended shelf life and product stability.
6. Students should be warned about the microbiological issues involved in sharing spoons or sampling from a communal container.
7. Trash bags or suitable containers should be provided for area cleanup as a large amount of packaging garbage can be generated.

15.3.3.2 Equipment
None.

15.3.3.3 Supplies
MREs (Meals, Ready to Eat), one for every two students.

Extra water may be needed to activate the heating elements although some recent versions have contained water packets.

Extra spoons, sample cups, and scissors or knives for opening the packets are required.

15.3.3.4 Suggested Grading
The questionnaire should be complete and easy to read. Many students will produce questionnaires that look quite similar, so it is difficult to evaluate whether work was an individual effort. One solution to this issue is to have students work in groups or pairs. This will generally result in a better final product. However, some students will contribute more than others in any group task.

15.3.4 Brief Exercise 4: Shelf Life Estimation

15.3.4.1 Notes and Keys to Successful Execution

1. This exercise is set up to be a visual test. No tasting is required although that is an option if the instructor wants to track a flavor change instead of a visual change. The reference by

Hough (2010) is recommended. It provides a thorough exposition of sensory shelf life testing and modeling.

2. Products. Bananas are a convenient product for a visual test of shelf life. You do not need to have the "consumers" actually eat the product since consumers "shop with their eyes." A bunch of bananas should be separated, placed in a normally lighted setting at room temperature and photographed at approximately the following times: 0, 2, 5, 8, 10, 12, and 15 days. Other intervals may be used depending upon how fast your fruits deteriorate in your climate/situation. The photographs can then be placed in a PowerPoint file or similar program for randomization of the stimulus order and labeling with three digit codes. The file is then shown to the students for their judgments. Bananas can be shown in duplicate or by simply rotating the orientation of the same fruit and showing it twice with different codes. It is not necessary to treat the replicates statistically, but they may just be averaged within each time block to get the overall proportion rejecting. Students should compute the values p, $1-p$, $p/(1-p)$, and $\ln(p/(1-p))$ for each time, which can be done simply in an Excel spreadsheet. The last value can then be plotted against time and against log time and fit by least squares methods to fit a logistic ("logit") function.

3. Probability papers. Preprinted probability papers and log probability papers mark off equal standard deviation units on the y-axis. They can be found in any college bookstore, statistics or engineering department, or downloaded from the web. Note that they are fitting to a normal distribution. Curvature in the fitted line (to time, not log time) would indicate the need for a lognormal approach.

4. Cut point alternatives. Rather than consumer rejection, the lab can also be done as if the students were a descriptive panel. In that case a suitable set of attributes should be developed such as degree of browning, extent of brown spotting, etc. The lab can then be further expanded to include both the consumer evaluation (that should be done first) and then a brief panel training exercise to orient the panel to the scales, their limits and reference standards, and then the descriptive ratings. The written exercise can then be extended to develop a descriptive cut point based on the pseudo-consumer data at 50 % failure time or some other measure. This can provide a good exercise in correlation and curve fitting to relate the descriptive and consumer data.

15.3.4.2 Equipment
Computer projector for PowerPoint slides or similar.
Digital camera for photographing fruit.

15.3.4.3 Supplies
Fresh fruit of good visual quality. Bananas are recommended for simplicity. The fruit should visually deteriorate over a period of about 2 weeks.

15.3.4.4 Further Background and Time Functions
It is common to describe a distribution of events by a location and shape parameter (e.g., a mean and standard deviation). If there is a bell-shaped symmetric distribution, two options are the simple normal distribution and a logistic distribution as shown below. The normal distribution function for a time-related distribution of event is given by

$$F(t) = \Phi\left(\frac{t-\mu}{\sigma}\right) \qquad (15.4)$$

Because many time-dependent processes are not normal or symmetric, a useful function is to model a lognormal distribution. In this case the distribution of events is right-skewed and becomes somewhat rarer as time progresses. The rejection function is then described by the following:

$$F(t) = \Phi\left(\frac{\ln(t)-\mu}{\sigma}\right) \qquad (15.5)$$

where Φ is the cumulative normal distribution, μ is the mean, and σ is the standard deviation, modeled as a function of time. Another common choice is the logistic function, which was originally used to model population growth that would approach some natural limit such as microbial growth:

$$F(t) = \frac{1}{1+e^{-1}} = \frac{e^1}{1+e^1} \qquad (15.6)$$

This relationship is often modeled by its logistic regression form, in which a linear function of the odds ratio is used to fit slope and intercept parameters:

$$\ln\left(\frac{p}{1-p}\right) = b_0 + b_1 t \qquad (15.7)$$

where p is the probability of rejection $[p = F(t)]$ and b_0 and b_1 are the parameters to be fit. In some cases, $\log(t)$ should be substituted for t.

Part IV

Statistical Problem Sets
for Sensory Evaluation

Sample Problem Sets for Statistics

16

16.1 Exercise 1: Means, Standard Deviations, and Standard Errors

Do not use a statistical package. Show your calculations. No calculations = no grade. Do not use Excel statistical functions. You can check your summations in Excel or a similar spreadsheet program if you wish.

We are interested in optimizing the spice level from pepper heat for a snack chip. The two alternative formulations are rated for perceived oral "burn" intensity by two mini-panels of 15 persons each chosen randomly from lab personnel. The ratings are given using the method of magnitude estimation relative to a standard and are shown below. The goal (unknown to the panelists) is to match the standard item, which is given a rating of "20" on this scale of 0–100 points.

1. For each group (product), compute the mean, the standard deviation, the standard error, and the median of the data.
2. Next, log transform the data and calculate the geometric mean. The geometric mean is the Nth root of the product of N items. In practice, it is calculated by first, taking logs, then, averaging the logs, and then, taking the antilog of that average.

3. Discuss as follows:
   (a) What would you conclude about the spice level? Does it appear to be close to the target/standard rating of 20?
   (b) What do you notice about the data sets with regard to symmetry and outliers (i.e., do they look like a bell-curve distribution)? An outlier is a data point that is very different from most of the others.
   (c) How did the geometric mean compare to the mean and the median?
   (d) What factors might be important in selecting panelists for this test?

16.2 Exercise 2: Binomial-Based Statistics for Discrimination Tests

An ABX test is conducted to determine the amount of added diacetyl that might be detected in cottage cheese by some panelists. Two tests are conducted. In both tests the control sample has no added diacetyl (but probably some small amount from the starter culture or the fermentation process). Test sample 389 has 1 ppm added diacetyl. Test sample 456 has 2 ppm added diacetyl.

H.T. Lawless, *Laboratory Exercises for Sensory Evaluation*, Food Science Text Series 2,
DOI 10.1007/978-1-4614-5713-8_16, © Springer Science+Business Media New York 2013

In the first ABX test with test sample 389, 36 consumers participate. 24 match the test sample to the correct reference item.

In the second ABX test with test sample 456, 15 consumers participate. 10 match the sample to the correct reference item.

1. What are the correct proportion and the z-score and p-value for each test?
2. Which test, if either, is statistically significant? Recall that the test is one-tailed and the critical z-value is 1.645.
3. What are the estimated *proportions* of discriminators in each test after you apply the correction for guessing (Abbott's formula)?
4. The test with 15 subjects was then replicated (same people tested again), and 13 out of 15 matched correctly. Why do you think this might have changed? What is the exact probability of a result of 13 or more out of 15 matching to the correct reference? (Hint: Show the binomial expansion for the needed terms and sum the probabilities; see Lawless and Heymann (2010) for an example of a binomial expansion.)
5. What was the null hypothesis? What was the alternative hypothesis? (Be sure to differentiate between sample proportions and population proportions!)

Useful equations are

$$z = \frac{\left(P_{obs} - P_{chance}\right) - \left(\frac{1}{2N}\right)}{\sqrt{\frac{pq}{N}}} \quad (16.1)$$

$$P_D = \frac{P_{obs} - P_{chance}}{1 - P_{chance}} \quad (16.2)$$

$$P(x) = \frac{N!}{X!(N-X)!}\left(p_{chance}\right)^X \left(1 - p_{chance}\right)^{N-x} \quad (16.3)$$

where P_{obs} is the proportion answering correctly, P_{chance} is the chance probability level, P_D is the proportion of discriminators, $P(x)$ is the probability of X occurrences out of N total trials in a binomial situation.

16.3 Exercise 3: The *t*-Tests

A group of panelists rates two products on a single 10-point category scale for the intensity of flavor (1=low, 10=high). Perform a dependent-group ("paired") t-test to see if there's any difference. Then, perform an independent-group t-test, as if the products had been given to two different panels and see if there's still a difference.

1. Compute: Means
   (a) Standard deviations
   (b) Standard errors of the mean
   (c) t-Value as paired test
   (d) t-Value as if independent groups
   Show: Calculations for both t-tests (paired t and independent groups t)
   Answer the following questions:
2. Which product received higher ratings?
3. Which t-test appeared to be more sensitive? Why? What is the advantage of having panelists evaluate both products? Why do the two t-tests give different answers?
4. What were your null and alternative hypotheses?

16.4 Exercise 4: Simple Correlation

A vegetable scientist would like to know the relationships among instrumentally measured sugar content and the perceived sweetness and starchy texture of frozen corn. A number of samples are collected at different times of the summer and submitted to both an instrumental analysis and a descriptive panel. Mean values are calculated for each "batch" and are given in the sample data set.

1. Find the simple correlation ("Pearson product–moment correlation," usually symbolized by the letter r) between each pair of measurements.

2. Are these values of r statistically significant (i.e., is the correlation greater than zero in absolute magnitude)? You can use a simple t-statistic on this where N is the number of pairs of observations and there are $N-2$ degrees of freedom.
3. What conclusions or recommendations can you draw for the client (veggie scientist)?
Useful Equations:

$$r = \frac{\sum XY - \left(\dfrac{\sum x \sum y}{N}\right)}{\sqrt{\left[\sum X^2 - \dfrac{(\sum X)^2}{N}\right]\left[\sum Y^2 - \dfrac{(\sum Y)^2}{N}\right]}}$$ (16.4)

$$t = r\sqrt{\frac{N-2}{1-r^2}}$$ (16.5)

16.5 Exercise 5: One and Two-Way ANOVA

Consult statistical Appendix C in Lawless and Heymann (2010) for additional help and worked examples.

A producer of chocolate milk desired to compare sweetness levels in their product ("own" brand) vs. a local competitor and a national brand made from a powdered mix. Each milk was rated on a nine-point scale for sweetness intensity as shown in the sample data set. 12 panelists rated all three samples. Data were coded as a score from 1 to 9.

1. Perform a simple one-way ANOVA, as if three separate groups of people had evaluated each milk. Show calculations and an ANOVA table (SS, df, MS, F). What is the critical F-value for significance? Are there any significant product differences?
2. Perform a two-way ("repeated measures") ANOVA that partitions out the panelist variation from error, treating the data as one group of 12 panelists where each panelist tried each milk. Show calculations and an ANOVA table.

What is the critical F-value for significance? Are there any significant product differences?
3. Which ANOVA uncovered the bigger difference? (if found)

16.6 Exercise 6: Planned Comparisons of Means Following ANOVA

A test is conducted with 14 trained panelists evaluating three products for flavor strength. The means are shown below.
Product means

Product A	Product B	Product C
6.55	5.36	4.33

A two-way analysis of variance is conducted. The panelists SS and MS are not needed for this exercise, and so they are not shown. Remember to subtract the panelist df from the total in computing the error.

Source	SS	df	MS	F
Products		2	31.2	
Error	42.16			(N/A)

1. Fill in the four missing blanks for the ANOVA table (using the information provided). Hint: You do not need to calculate the squared totals to get the product SS.
2. Is there evidence for a significant difference among the product means? (Explain why or why not.)
3. Conduct LSD tests on the three product means. Show your work. Which pairs are different?
4. Conduct Duncan tests on the three product means. Show your work. Which pairs are different?

The repeated measures ANOVA are found in the complete block design section of your Appendix C in pages 510–512. Examples of the LSD test and Duncan test are found in pages 513–514 of Appendix C.

16.7 Exercise 7: Rank Order Tests

Three flavors are ranked by 15 people in a laboratory acceptance panel to screen possible flavors for their appropriateness in a cottage cheese — pineapple, garlic, and mint. The data set shows the ranks; a rank of one is the most preferred.

1. Analyze these data by Kramer's rank sum test or the tables of Newell and MacFarlane (see page 563 of Lawless and Heymann (2010)) and by the Friedman nonparametric "analysis of variance."

2. Compare the rankings of all three pairs of products using the LSD test formula shown below. Give significance of results for all tests.

3. Perform a sign test on each pair. Report probability and significance. Remember this should be two-tailed because they are preference rankings.

Useful equations:

The existence of a difference is tested in the Friedman "analysis of variance on ranks" as a chi square variable with $J-1$ degrees of freedom

$$\chi^2 = \left\{ \frac{12}{[K(J)(J+1)]} \left[\sum_{j=1}^{J} T_j^2 \right] \right\} - 3K(J+1) \quad (15.6)$$

for a table of K rows (judges) and J columns (products) and where T_j are the column totals.

To compare individual products, an LSD test for rank sums requires minimum differences equal to

$$LSD = 1.96 \sqrt{\frac{K(J)(J+1)}{6}} \quad (15.7)$$

for J items ranked by K panelists.

16.8 Notes to Instructors

The first seven problem sets are designed to be done without a statistical program, i.e., with a hand calculator. The instructor should decide whether there is value in having students do such calculations by hand or whether a statistical package can be used. One philosophy is that the students will learn the procedures better and have a fuller understanding if they do the calculations themselves. A compromise is to allow the use of a spreadsheet program for summation, squaring, and other simple tasks. This can help eliminate keying errors. Students may then be required to show the additional calculations and how the obtained sums or squared values are used in the statistical equations.

Students will come to a sensory evaluation class with different statistical backgrounds. Some courses require a stat course as a prerequisite, but unless the students have taken one recently, they may not have retained those skill sets. The problems are designed to insure a basic level of statistical proficiency. Depending upon the level of the course, further more advanced topics may be covered. Worked examples of these exercises can be found in the statistical appendix sections of Lawless and Heymann (2010).

16.9 For Further Reading

Lawless HT, Heymann H (2010) Sensory evaluation of foods, principles and practices, 2nd ed., Springer Science+Business, New York

16.10 Appendix: Sample Data Sets and Open Data Tables

Exercise 1: Sample Data Set

Group 1, Product 1	Ratings	Group 2, Product 2	Ratings
judge 1	11	judge 1	16
judge 2	16	judge 2	18
judge 3	23	judge 3	22
judge 4	35	judge 4	38
judge 5	25	judge 5	17
judge 6	18	judge 6	63
judge 7	22	judge 7	39
judge 8	18	judge 8	15
judge 9	22	judge 9	16
judge 10	24	judge 10	23
judge 11	22	judge 11	22
judge 12	29	judge 12	75
judge 13	15	judge 13	22
judge 14	14	judge 14	10
judge 15	16	judge 15	12

You can use this open table to show your calculations (it is helpful to see your numbers organized like this):

	Group 1			Group 2		
Judge Number	Data ("X")	X^2	Log X	Data ("Y")	Y^2	Log Y
1						
2						
3						
4						
5						
6						
7						
8						
9						
10						
11						
12						
13						
14						
15						
SUM						

Exercise 3, Sample data set for t-tests

PANELIST	PRODUCT A	PRODUCT B
1	3	5
3	4	6
4	5	7
5	6	9
6	5	4
7	6	8
8	7	5
9	7	8
10	7	6
11	7	9
12	3	7
13	5	8
14	7	9
15	8	9

Open table for calculations

	Product A		Product B		Difference, D	
Judge Number	Data ("X")	X^2	Data ("Y")	Y^2	X-Y	D^2
1						
2						
3						
4						
5						
6						
7						
8						
9						
10						
11						
12						
13						
14						
15						
SUM						

Exercise4: Simple correlation

Sample data set:

Sweetness score	Sugar (%w/v)	Starchiness score
7	8.0	2
4	5.0	5
6	6.0	3
7	8.1	2.5
2	3.1	6
1	3.2	8
2	3.3	7
6	7.0	4
5	6.1	1
8	7.2	1.5

Open data table (duplicate as necessary)

	Variable 1		Variable 2		Product
Judge Number	Data ("X")	X^2	Data ("Y")	Y^2	X*Y
1					
2					
3					
4					
5					
6					
7					
8					
9					
10					
SUM					

Exercise 5: ANOVA

Sample data set:

PANELIST	NATIONAL COMPETITOR	LOCAL COMPETITOR	"OWN" BRAND
1	7	4	3
2	9	6	5
3	7	3	6
4	4	2	6
5	9	7	5
6	8	4	6
7	4	2	1
8	3	1	2
9	6	4	2
10	3	2	1
11	7	5	4
12	8	6	4

Open data table:

Judge Number	Variable 1 Data ("X")	X^2	Variable 2 Data ("Y")	Y^2	Variable 3 Data ("Z")	Z^2	Sum (rows) (X+Y+Z))	Sum^2
1								
2								
3								
4								
5								
6								
7								
8								
9								
10								
11								
12								
Sum of columns								?
Sum^2		?		?		?		?

? indicates that cell or sum will probably not be needed. So check the formulas before you fill it in.

Exercise 7: Rank order tests

Sample data set:

Panelist	Pineapple	Garlic	Mint
1	1	2	3
2	2	1	3
3	3	2	1
4	1	2	3
5	1	2	3
6	1	2	3
7	1	2	3
8	3	1	2
9	1	3	2
10	1	2	3
11	2	1	3
12	3	1	2
13	1	3	2
14	1	2	3
15	1	3	2

Index

H.T. Lawless, *Laboratory Exercises for Sensory Evaluation*, Food Science Text Series 2,
DOI 10.1007/978-1-4614-5713-8, © Springer Science+Business Media New York 2013

CPSIA information can be obtained
at www.ICGtesting.com
Printed in the USA
LVHW061523150820
663276LV00004B/300